Leckie
the education publisher
for Scotland

Primary Maths for Scotland

2nd Level Maths
Assessment Pack

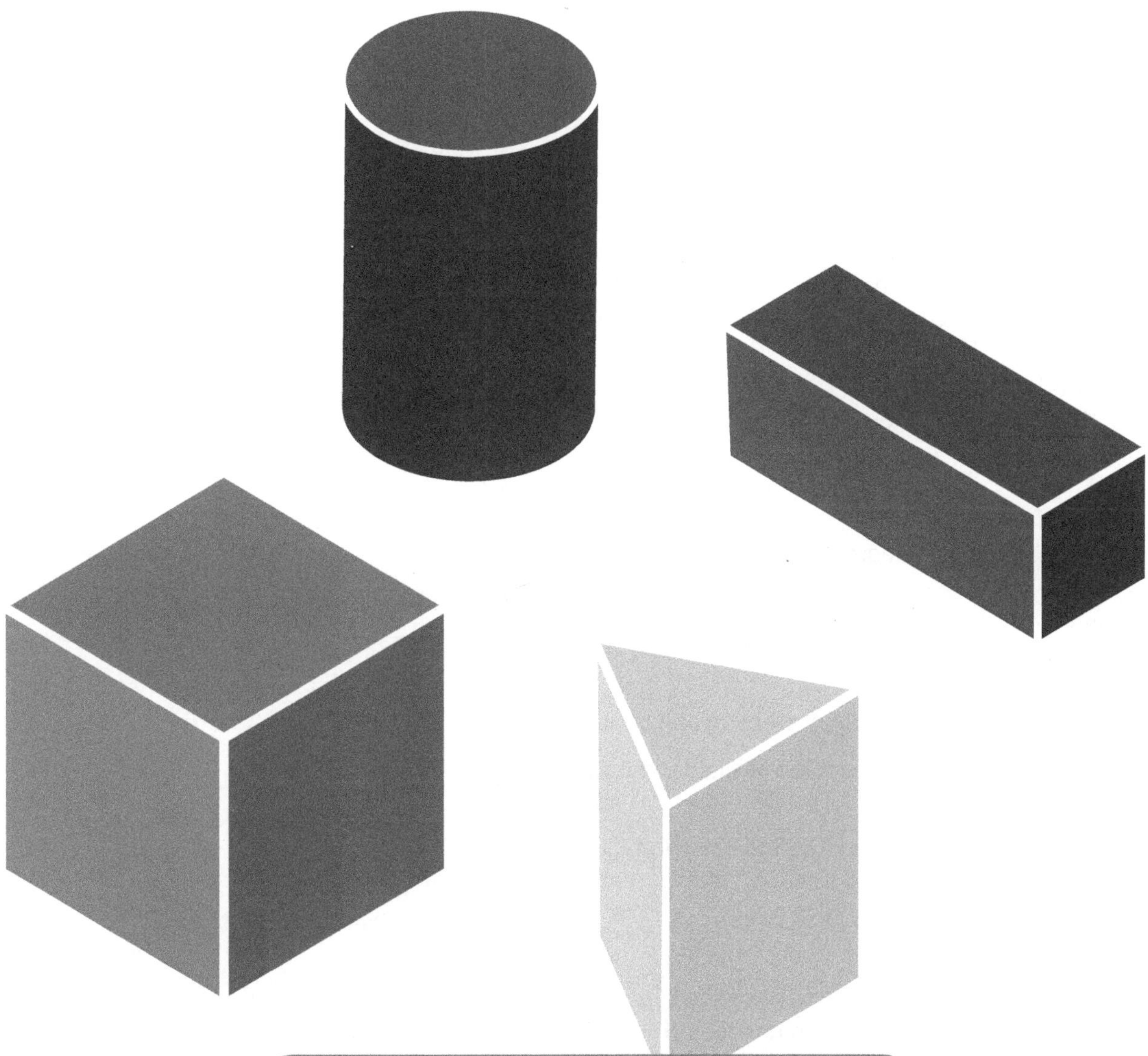

Authors: Linda Lapere and Scott Morrow
Assessment Consultant: Carol Lyon
Series Editor: Craig Lowther

001/27012021

10 9 8 7 6 5 4 3 2 1

ISBN 9780008392482

Published by
Leckie & Leckie Ltd
An imprint of HarperCollinsPublishers
Westerhill Road, Bishopbriggs, Glasgow, G64 2QT
T: 0844 576 8126 F: 0844 576 8131
leckieandleckie@harpercollins.co.uk
www.leckieandleckie.co.uk

HarperCollins Publishers
1st Floor, Watermarque Building,
Ringsend Road, Dublin 4, Ireland

Publisher: Fiona McGlade
Project manager: Rachel Allegro

Special thanks to
Copy editor: Mitch Fitton
Layout and illustration: Jouve
Proofreaders: Louise Robb, Deborah Dobson and Lauren Reid

A CIP Catalogue record for this book is available from the British Library

Printed and bound by CPI Group (UK) Ltd, Croydon, CR0 4YY

Acknowledgements
Images © Shutterstock

Whilst every effort has been made to trace the copyright holders, in cases where this has been unsuccessful, or if any have advertently been overlooked, the Publishers would gladly receive any information enabling them to rectify any error or omission at the first opportunity.

Contents

Find all the downloadable resources – including Yearly Progress Checks, End of Level Assessments and record sheets – here: collins.co.uk/primarymathsforscotland

Introduction

This Assessment Pack includes high quality assessments for Numeracy and Mathematics. It can be used to support ongoing professional judgement of pupils' progress and next steps within the Scottish Curriculum for Excellence (CfE) Second Level. Assessment Packs for Early and First Levels are also available, to ensure consistency from Preschool to P7.

As stated in Building the Curriculum 5 – A Framework for Assessment, we at Leckie understand that ongoing assessment and pupil feedback is vital to, '…support learning and promote learner engagement resulting in greater breadth and depth in learning, including a greater focus on the secure development of knowledge, understanding and skills', Scottish Government (2011).

In this introduction, you will find:

1. Guidance for assessment
2. Introduction to the assessments supplied in this pack
3. Support for record keeping
4. Tables showing coverage of the Experiences and Outcomes and Benchmarks

Guidance for assessment

Ongoing formative assessment is an essential part of the learning process. Black and Wiliam (2009) listed five key principles of learning:

1. Clarifying and sharing learning intentions and success criteria;
2. Engineering effective classroom discussions and other learning tasks that elicit evidence of student understanding;
3. Providing feedback that moves learners forward;
4. Activating students as instructional resources for one another; and
5. Activating students as the owners of their own learning.

The Leckie Primary Maths for Scotland Textbooks and Teacher Guides were designed and written with these principles at the forefront, to support teachers to plan and deliver excellent mathematical learning. Both resources create opportunities for: diagnostic questions that assess prior learning; observation of and feedback on day-to-day learning (including outdoor learning); leading and listening to learning conversations.

The Leckie Primary Maths for Scotland Assessment Packs now add to this by providing rich, diagnostic tasks and activities to help the educator track and monitor progress and achievement. The materials contained within the packs are both summative and formative in nature. Not only do they provide important information about the pupil's current knowledge and skills (summative information), they also help diagnose misconceptions and identify next steps for both the educator and pupil (formative information).

The Leckie Teacher Guides contain advice and support that will help the educator address common misconceptions that children experience as they learn and develop mathematical skills, thus helping to ensure smooth and continuous progress.

Throughout the learning, teaching and assessment process, feedback should be used to ensure the pupil can answer these questions:

- Where am I going?
- How am I going there?
- Where to next?

Hattie, J.A.C. (2012)

Primary Maths for Scotland Second Level Assessment Pack

The Primary Maths for Scotland Second Level Assessment Pack contains two types of assessment:

1. Yearly Progress Checks
2. End of Level Assessments

Yearly Progress Checks (YPC)

These checks provide full coverage of the CfE Second Level Experiences and Outcomes. Each progress check assesses mental agility, procedural fluency and conceptual understanding of the key numeracy and mathematical ideas appropriate to this level.

When to use the Yearly Progress Checks

There are three Yearly Progress Checks included in this pack. Yearly Progress Check 2A is designed to be used at the end of Textbook 2A. Yearly Progress Check 2B is designed to be used at the end of Textbook 2B. Yearly Progress Check 2C is designed to be used at the end of Second Level which, for almost all pupils, will be at the end of P7.

How to use the Yearly Progress Checks

Yearly Progress Checks can be completed in groups or by the whole class. The questions have been arranged in the order of the Experiences and Outcomes for ease of tracking. However, teachers should use their professional judgement in deciding how the assessment is administered and questions grouped together.

The Yearly Progress Checks are a blend of questions where the pupil will need to write, draw, choose or say their answer. To help diagnose understanding and misconceptions, we recommend that pupils are encouraged to explain their thinking and show their understanding using manipulatives and written answers. If reading or writing is a barrier to the pupil sharing their answer, then we recommend that the questions are read out and the pupil explains their answer and thinking orally.

The Yearly Progress Checks include a mix of bare number problems, questions using pictures and diagrams and contextualised problems. Throughout the assessment, the pupil is encouraged to explain their thinking to help you to understand the strategy they are using. This is important as the same question can be solved in many, increasingly sophisticated ways.

At times, the pupil will become stuck on a question; this is to be expected. It may be appropriate for you to read or restate the question, perhaps in a context with which the pupil is more familiar. Be mindful, however, not to make the question easier or to give away a strategy that the pupil can use. Wherever a pupil struggles, this is evidence of next steps in learning.

The answers the pupil provides will be an indication of the mathematical and numerical facts that they have committed to memory and the strategies and skills they are using to solve problems.

Misconceptions will also be highlighted by the responses given. The Primary Maths for Scotland Teacher Guides list possible misconceptions by topic and provide help and suggested tasks for the pupil to undertake. By comparing the data collected from a group or class it will also be possible to check how well a particular topic was learned and whether the issue lies with the individual or the group.

Using the Marking Guidance

The Marking Guidance provided for the Yearly Progress Checks includes a reference to the appropriate Experience(s) and Outcome(s) linked to each question. The 'Notes' section explains what to expect of a pupil who is 'on track' within the level and provides advice on 'further learning required' for the pupil who has struggled or could not answer the question. Where further learning is required, the Leckie Teacher Guide describes common misconceptions and offers advice on appropriate learning activities to help address these.

Question:

29. a There are 100 packets of 15 felt pens in a box. Finlay says that means there are 150 felt pens altogether, Amman thinks 1500. Who is correct? Explain your thinking.

b Finlay thought of a number. He multiplied it by 10 and then by 10 again. He now has 800. What number did he start with? Explain your answer.

Marking Guidance:

Topic–benchmarks / Es & Os	Question	Answer	Marking guidance
MNU 2-02a MNU 2-03a	29	a Amman is correct. $15 \times 100 = 1500$. b Finlay started with 8. $8 \times 10 \times 10 = 800$	**On track** • Response to question a shows that the pupil understands what happens to a number when we multiply it by 100. • Can use their knowledge of multiplication as the inverse of division to complete question b, i.e. by working backwards and dividing 800 by 10 then by 10 again to find the unknown. **Further learning required** • May respond or calculate inaccurately. • May justify their response using an incorrect or ineffective strategy, e.g. 'Add two zeros when multiplying by 100'. • May be unable to work backwards to solve b. May answer 80 000 because they 'see' $800 \times 10 \times 10$.

End of Level Assessments (EOL)

These assessments are contextualised, holistic assessments that blend together topics from across the Mathematics and Numeracy Curriculum. There are several questions per assessment that can be used to provide a broad picture of the pupil's mastery of the national standards set out in the Benchmarks.

The questions will be in new and unfamiliar contexts to the pupil. This will increase the challenge for them as they have to both understand the context and work out what strategy, skills and knowledge to use. As with the Yearly Progress Checks, the pupil is encouraged to show their thinking to help you understand the strategies they are using and identify any misconceptions.

Introduction

When to use the End of Level Assessments

As the name suggests, the End of Level Assessments should be used when you judge that a pupil or group of pupils has mastered the Second Level Mathematics Curriculum, or to ascertain whether they have achieved mastery in instances where you are unsure. This will generally be at the end of P7 if the pupil is meeting age-related expectations.

How to use the End of Level Assessments

The End of Level Assessments are a blend of activities and questions linked to a context. Being able to answer problems set in new and unfamiliar contexts demonstrates mastery of the mathematics set out in the curriculum. Sometimes the context can become a significant barrier to progress. If this is the case then you may wish to stop the assessment and explore the context through storytelling, watching videos, playing games, discussion and linking the assessment activity to previous learning.

The assessments can be completed in groups or by individuals. When pupils complete the assessment tasks in groups, they will discuss their ideas and allow the teacher to observe the skills and knowledge they have mastered. Throughout the assessment it is permitted for the pupils to make informal jottings, use concrete materials, create paper models or templates, or draw pictures or diagrams.

Using the Marking Guidance

The Marking Guidance for the End of Level Assessments includes the Experiences and Outcomes for each question part, the question, a description of the answer that tells you the pupil is 'on track' and a 'review' section that describes what you can expect a pupil to do if they have not mastered the concept assessed. The Leckie Teacher Guide has further guidance on how to support pupils who were unable to engage in the assessment tasks, to help move their learning forwards.

Number and number processes	Having determined which calculations are needed, I can solve problems involving whole numbers using a range of methods, sharing my approaches and solutions with others. MNU 2-03a I have explored the contexts in which problems involving decimal fractions occur and can solve related problems using a variety of methods. MNU 2-03b	• Adds and subtracts whole numbers and decimal fractions to two decimal places, within the number range 0 to 1 000 000. • Multiplies and divides whole numbers by multiples of 10, 100, 1000 • Divides whole numbers and decimal fractions to two decimal places, by a single digit, including answers expressed as decimal fractions, for example $43 \div 5 = 8.6$

Record Keeping

Yearly Progress Checks Record Keeping

There are record sheets for you to use to track the progress of the pupils against each question. We suggest that you 'mark' each pupil as they complete their Second Level Yearly Progress Check. We suggest the ABC coding described below but this can easily be replaced with green/amber/red to provide a quick visual check:

A. Chose an appropriate strategy/method and used it correctly (green)

B. Chose an appropriate strategy/method but used it incorrectly or made an error in calculation (amber)

C. Chose an inappropriate strategy/method or did not attempt the question (red)

End of Level Assessment Record Keeping

We have provided record sheets for you to use as the pupils complete the End of Level Assessments and suggest using the coding:

- O. On track
- R. Review

Again, a simple colour coding system can be used to provide a quick visual check, e.g. using green for 'on track' and red for 'review'.

It is important to note that a pupil does not need to successfully complete every task contained within the Leckie Primary Maths for Scotland Assessment Pack in order to achieve CfE Second Level. This decision will be based on the teacher's professional judgement as they observe their pupils engaged in their day-to-day learning. Nevertheless, we hope the record sheets included will provide a simple reference to where pupils have mastery of key Numeracy and Mathematics concepts and where they require further support.

All record sheets can be found online at: collins.co.uk/primarymathsforscotland

References

Black P. and Wiliam D. (2009) Developing the theory of formative assessment. *Educational Assessment, Evaluation and Accountability*, 21(1), 5–31

Hattie, J.A.C. (2012) *Visible Learning for Teachers*. London. Routledge

Scottish Government (2011) *Curriculum for Excellence. Building the Curriculum 5 – A Framework for Assessment*. Available online https://www.education.gov.scot/Documents/btc5-framework.pdf [accessed on 01/10/2020]

Coverage of the CfE Numeracy and Mathematics Experiences and Outcomes and Benchmarks

Curriculum organisers	Experiences and Outcomes	Benchmarks to support practitioners' professional judgement of achievement of a level	Coverage in Second Level Assessment Pack
Estimation and rounding	I can use my knowledge of rounding to routinely estimate the answer to a problem then, after calculating, decide if my answer is reasonable, sharing my solution with others. MNU 2-01a	• Rounds whole numbers to the nearest 1000, 10 000 and 100 000. • Rounds decimal fractions to the nearest whole number, to one decimal place and two decimal places. • Applies knowledge of rounding to give an estimate to a calculation appropriate to the context.	YPC 2A, 2B, 2C EOL 1, 2
Number and number processes	I have extended the range of whole numbers I can work with and having explored how decimal fractions are constructed, can explain the link between a digit, its place and its value. MNU 2-02a	• Reads, writes and orders whole numbers to 1 000 000, starting from any number in the sequence. • Explains the link between a digit, its place and its value for whole numbers to 1 000 000. • Reads, writes and orders sets of decimal fractions to three decimal places. • Explains the link between a digit, its place and its value for numbers to three decimal places. • Partitions a wide range of whole numbers and decimal fractions to three decimal places; for example, 3.6 = 3 ones and 6 tenths = 36 tenths.	YPC 2A, 2B, 2C EOL 6
	Having determined which calculations are needed, I can solve problems involving whole numbers using a range of methods, sharing my approaches and solutions with others. MNU 2-03a	• Adds and subtracts multiples of 10, 100 and 1000 to and from whole numbers and decimal fractions to two decimal places. • Adds and subtracts whole numbers and decimal fractions to two decimal places, within the number range 0 to 1 000 000. • Uses multiplication and division facts to the 10th multiplication table. • Multiplies and divides whole numbers by multiples of 10, 100 and 1000. • Multiplies and divides decimal fractions to two decimal places by 10, 100 and 1000.	YPC 2A, 2B, 2C EOL 1, 2, 6

Curriculum organisers	Experiences and Outcomes	Benchmarks to support practitioners' professional judgement of achievement of a level	Coverage in Second Level Assessment Pack
	I have explored the contexts in which problems involving decimal fractions occur and can solve related problems using a variety of methods. MNU 2-03b	• Multiplies whole numbers by two digit numbers. • Multiplies decimal fractions to two decimal places by a single digit. • Divides whole numbers and decimal fractions to two decimal places, by a single digit, including answers expressed as decimal fractions; for example, $43 \div 5 = 8.6$.	YPC 2A, 2B, 2C EOL 1, 2, 3, 6
	Having explored the need for rules for the order of operations in number calculations, I can apply them correctly when solving simple problems. MTH 2-03c	• Applies the correct order of operations in number calculations when solving multi-step problems.	YPC 2A, 2B, 2C
	I can show my understanding of how the number line extends to include numbers less than zero and have investigated how these numbers occur and are used. MNU 2-04a	• Identifies familiar contexts in which negative numbers are used. • Orders numbers less than zero and locates them on a number line.	YPC 2A, 2B, 2C
Multiples, factors and primes	Having explored the patterns and relationships in multiplication and division, I can investigate and identify the multiples and factors of numbers. MTH 2-05a	• Identifies multiples and factors of whole numbers and applies knowledge and understanding of these when solving relevant problems in number, money and measurement.	YPC 2A, 2B, 2C EOL 3

Curriculum organisers	Experiences and Outcomes	Benchmarks to support practitioners' professional judgement of achievement of a level	Coverage in Second Level Assessment Pack
Fractions, decimal fractions and percentages	I have investigated the everyday contexts in which simple fractions, percentages or decimal fractions are used and can carry out the necessary calculations to solve related problems. MNU 2-07a	• Uses knowledge of equivalent forms of common fractions, decimal fractions and percentages; for example, $\frac{3}{4} = 0.75 = 75\%$, to solve problems. • Calculates simple percentages of a quantity, and uses this knowledge to solve problems in everyday contexts; for example, calculates the sale price of an item with a discount of 15%. • Calculates simple fractions of a quantity and uses this knowledge to solve problems; for example, find $\frac{3}{5}$ of 60.	YPC 2A, 2B, 2C EOL 1, 4
	I can show the equivalent forms of simple fractions, decimal fractions and percentages, and can choose my preferred form when solving a problem, explaining my choice of method. MNU 2-07b	• Creates equivalent fractions and uses this knowledge to put a set of most commonly used fractions in order.	YPC 2A, 2B, 2C EOL 6
	I have investigated how a set of equivalent fractions can be created, understanding the meaning of simplest form, and can apply my knowledge to compare and order the most commonly used fractions. MTH 2-07c	• Expresses fractions in their simplest form.	YPC 2A, 2B, 2C EOL 6

Curriculum organisers	Experiences and Outcomes	Benchmarks to support practitioners' professional judgement of achievement of a level	Coverage in Second Level Assessment Pack
Money	I can manage money, compare costs from different retailers, and determine what I can afford to buy. MNU 2-09a	• Carries out money calculations involving the four operations. • Compares costs and determines affordability within a given budget.	YPC 2A, 2B, 2C
	I understand the costs, benefits and risks of using bank cards to purchase goods or obtain cash and realise that budgeting is important. MNU 2-09b	• Demonstrates understanding of the benefits and risks of using bank cards and digital technologies.	YPC 2B, 2C EOL 2, 4
	I can use the terms profit and loss in buying and selling activities and can make simple calculations for this. MNU 2-09c	• Calculates profit and loss accurately; for example, when working with a budget for an enterprise activity.	YPC 2A, 2B, 2C EOL 5
Time	I can use and interpret electronic and paper-based timetables and schedules to plan events and activities, and make time calculations as part of my planning. MNU 2-10a	• Reads and records time in both 12-hour and 24-hour notation and converts between the two. • Knows the relationships between commonly used units of time and carries out simple conversion calculations; for example, changes $1\frac{3}{4}$ hours into minutes. • Uses and interprets a range of electronic and paper-based timetables and calendars to plan events or activities and solve real-life problems. • Calculates durations of activities and events including situations bridging across several hours and parts of hours using both 12-hour clock and 24-hour notation.	YPC 2A, 2B, 2C EOL 1, 4, 6

Curriculum organisers	Experiences and Outcomes	Benchmarks to support practitioners' professional judgement of achievement of a level	Coverage in Second Level Assessment Pack
	I can carry out practical tasks and investigations involving timed events and can explain which unit of time would be most appropriate to use. MNU 2-10b	• Chooses the most appropriate timing device in practical situations and records using relevant units, including hundredths of a second. • Selects the most appropriate unit of time for a given task and justifies choice.	YPC 2A, 2B, 2C
	Using simple time periods, I can give a good estimate of how long a journey should take, based on my knowledge of the link between time, speed and distance. MNU 2-10c	• Estimates the duration of a journey based on knowledge of the link between speed, distance and time.	YPC 2C EOL 2
Measurement	I can use my knowledge of the sizes of familiar objects or places to assist me when making an estimate of measure. MNU 2-11a	• Uses the comparative size of familiar objects to make reasonable estimations of length, mass, area and capacity. • Estimates to the nearest appropriate unit, then measures accurately: length, height and distance in millimetres (mm), centimetres (cm), metres (m) and kilometres (km); mass in grams (g) and kilograms (kg); and capacity in millilitres (ml) and litres (l).	YPC 2A, 2B, 2C EOL 2
	I can explain how different methods can be used to find the perimeter and area of a simple 2D shape or volume of a simple 3D object. MNU 2-11c	• Calculates the perimeter of simple straight-sided 2D shapes in millimetres (mm), centimetres (cm) and metres (m). • Calculates the area of squares, rectangles and right-angled triangles in square millimetres (mm^2), square centimetres (cm^2) and square metres (m^2). • Calculates the volume of cubes and cuboids in cubic centimetres (cm^3) and cubic metres (m^3). • Draws squares and rectangles accurately with a given perimeter or area. • Demonstrates understanding of the conservation of measurement; for example, draw three different rectangles each with an area of 24 cm^2.	YPC 2A, 2B, 2C EOL 3, 5

Curriculum organisers	Experiences and Outcomes	Benchmarks to support practitioners' professional judgement of achievement of a level	Coverage in Second Level Assessment Pack
	I can use the common units of measure, convert between related units of the metric system and carry out calculations when solving problems. MNU 2-11b	• Converts between common units of measurement using decimal notation; for example, 550 cm = 5.5 m; 3.009 kg = 3009 g. • Chooses the most appropriate measuring device for a given task and carries out the required calculation, recording results in the correct unit. • Reads a variety of scales accurately. • Shows awareness of imperial units used in everyday life; for example, miles or stones.	YPC 2A, 2B, 2C EOL 2, 3, 5
Mathematics – its impact on the world, past, present and future	I have worked with others to explore, and present our findings on, how mathematics impacts on the world and the important part it has played in advances and inventions. MTH 2-12a	• Researches and presents examples of the impact mathematics has in the world of life and work. • Contributes to discussions and activities on the role of mathematics in the creation of important inventions, now and in the past.	YPC 2A, 2B, 2C
Patterns and relationships	Having explored more complex number sequences, including well-known named number patterns, I can explain the rule used to generate the sequence, and apply it to extend the pattern. MTH 2-13a	• Explains and uses a rule to extend well-known number sequences, including square numbers, triangular numbers and Fibonacci sequence. • Applies knowledge of multiples, square numbers and triangular numbers to generate number patterns.	YPC 2A, 2B, 2C EOL 3
Expressions and equations	I can apply my knowledge of number facts to solve problems where an unknown value is represented by a symbol or letter. MTH 2-15a	• Solves simple algebraic equations with one variable; for example, $a - 30 = 40$ and $4b = 20$.	YPC 2A, 2B, 2C EOL 3

Curriculum organisers	Experiences and Outcomes	Benchmarks to support practitioners' professional judgement of achievement of a level	Coverage in Second Level Assessment Pack
Properties of 2D shapes and 3D objects	Having explored a range of 3D objects and 2D shapes, I can use mathematical language to describe their properties, and through investigation can discuss where and why particular shapes are used in the environment. MTH 2-16a	• Describes 3D objects and 2D shapes using specific vocabulary including regular, irregular, diagonal, radius, diameter and circumference. Applies this knowledge to demonstrate understanding of the relationship between 3D objects and their nets. • Identifies and describes 3D objects and 2D shapes within the environment and explains why their properties match their function. • Knows that the radius is half of the diameter.	YPC 2A, 2B, 2C
	Through practical activities, I can show my understanding of the relationship between 3D objects and their nets. MTH 2-16b		YPC 2A, 2B, 2C EOL 3
	I can draw 2D shapes and make representations of 3D objects using an appropriate range of methods and efficient use of resources. MTH 2-16c	• Uses digital technologies and mathematical instruments to draw 2D shapes and make representations of 3D objects, understanding that not all parts of the 3D object can be seen.	YPC 2A, 2B, 2C
Angle, symmetry and transformation	I have investigated angles in the environment, and can discuss, describe and classify angles using appropriate mathematical vocabulary. MTH 2-17a	• Uses mathematical language including acute, obtuse, straight and reflex to describe and classify a range of angles identified within shapes in the environment.	YPC 2A, 2B, 2C EOL 4, 5

Curriculum organisers	Experiences and Outcomes	Benchmarks to support practitioners' professional judgement of achievement of a level	Coverage in Second Level Assessment Pack
	I can accurately measure and draw angles using appropriate equipment, applying my skills to problems in context. MTH 2-17b	• Measures and draws a range of angles to within ±2°. • Knows that complementary angles add up to 90° and supplementary angles add up to 180° and uses this knowledge to calculate missing angles.	YPC 2A, 2B, 2C EOL 6
	Through practical activities that include the use of technology, I have developed my understanding of the link between compass points and angles and can describe, follow and record directions, routes and journeys using appropriate vocabulary. MTH 2-17c	• Uses knowledge of the link between the eight compass points and angles to describe, follow and record directions.	YPC 2A, 2B, 2C
	Having investigated where, why and how scale is used and expressed, I can apply my understanding to interpret simple models, maps and plans. MTH 2-17d	• Interprets maps, models or plans with simple scales; for example, 1 cm to 2 km.	YPC 2A, 2B, 2C

Curriculum organisers	Experiences and Outcomes	Benchmarks to support practitioners' professional judgement of achievement of a level	Coverage in Second Level Assessment Pack
	I can use my knowledge of the coordinate system to plot and describe the location of a point on a grid. MTH 2-18a / MTH 3-18a	• Describes, plots and records the location of a point, in the first quadrant, using coordinate notation.	YPC 2A, 2B, 2C
	I can illustrate the lines of symmetry for a range of 2D shapes and apply my understanding to create and complete symmetrical pictures and patterns. MTH 2-19a / MTH 3-19a	• Identifies and illustrates line symmetry on a wide range of 2D shapes and applies this understanding to complete a range of symmetrical patterns, with and without the use of digital technologies.	YPC 2A, 2B, 2C
Data and analysis	Having discussed the variety of ways and range of media used to present data, I can interpret and draw conclusions from the information displayed, recognising that the presentation may be misleading. MNU 2-20a	• Analyses, interprets and draws conclusions from a variety of data. • Draws conclusions about the reliability of data, taking into account, for example, the author, the audience, the scale and sample size used.	YPC 2A, 2B, 2C EOL 1
	I have carried out investigations and surveys, devising and using a variety of methods to gather information and have worked with others to collate, organise and communicate the results in an appropriate way. MNU 2-20b	• Devises ways of collecting data in the most suitable way for the given task. • Collects, organises and displays data accurately in a variety of ways including through the use of digital technologies; for example, creating surveys, tables, bar graphs, line graphs, frequency tables, simple pie charts and spreadsheets.	YPC 2A, 2B, 2C

Curriculum organisers	Experiences and Outcomes	Benchmarks to support practitioners' professional judgement of achievement of a level	Coverage in Second Level Assessment Pack
	I can display data in a clear way using a suitable scale, by choosing appropriately from an extended range of tables, charts, diagrams and graphs, making effective use of technology. MTH 2-21a / MTH 3-21a	• Displays data appropriately, making effective use of technology and chooses a suitable scale when creating graphs.	YPC 2A, 2B, 2C
Ideas of chance and uncertainty	I can conduct simple experiments involving chance and communicate my predictions and findings using the vocabulary of probability. MNU 2-22a	• Uses the language of probability accurately to describe the likelihood of simple events occurring; for example, equal chance; fifty-fifty; one in two; two in three; percentage chance; and $\frac{1}{6}$. • Plans and carries out simple experiments involving chance with repeated trials; for example, 'what is the probability of throwing a six if you throw a die fifty times?'. • Uses data to predict the outcome of a simple experiment.	YPC 2A, 2B, 2C EOL 4

Yearly progress check 2A

Resources

Resources provided online:

Digit cards

Resources needed and available in the setting:

Rulers • Protractors • Isometric dotted paper • Squared paper

Questions

1. What number is this?

   356

   a Write it in words. ______________________

   b What is 356 rounded to the nearest 10? ______________________

   c What is 356 rounded to the nearest 100? ______________________

2. Amman says that 647 rounded to the nearest 100 is 700.

   Finlay says that 647 rounded to the nearest 100 is 600.

   Who is correct? Explain your answer.

3. Write the numbers in the correct position on the number line:

   0·7 1 1·4 1·6

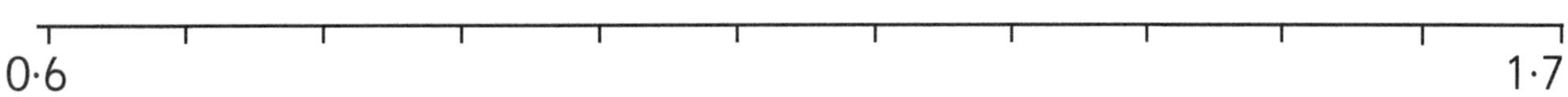

4. The table below shows the average temperatures in five cities in Scotland.

City	Average temperature in January
Stirling	4·9°c
Aberdeen	3·1°c
Perth	3·8°c
Edinburgh	5·2°c
Oban	2·9°c

Round each of the temperatures to the nearest whole number. What cities have the same average temperature for January after they have been rounded?

5. Estimate the answers to these additions by rounding to the nearest 10:

21 + 87 = ______ 45 + 67 = ______ 34 + 75 = ______

Calculate the exact answers. How reasonable were your estimates?

6. 6 0 1 9

Finlay thinks the smallest 4-digit number possible using these cards is 0169. Isla thinks it is 1069. Who do you agree with and why?

7.

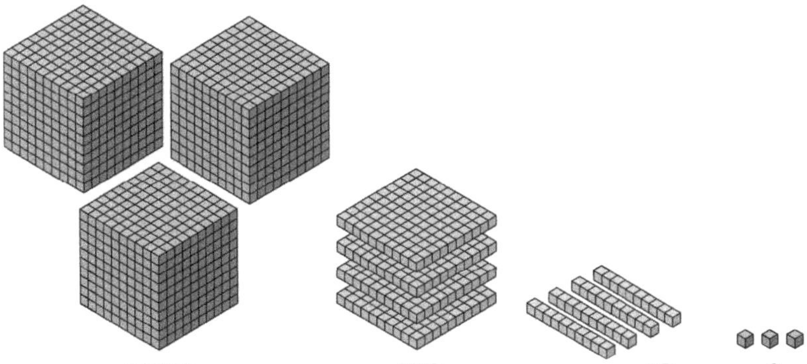

a What number does this represent? Write this number in numerals and words.

____

b Isla is thinking of a number. It has a 7 in the ones place, a 4 in the thousands place, a place holder in the hundreds place and a 2 in the tens place. Write down Isla's number.

____

8. Fill in the empty boxes.

The number 367 can be partitioned into [300] and [60] and [7].

367 can also be partitioned into [200] and [ ] and [17].

367 can also be partitioned into [ ] and [ ] and [ ].

9. Use these digit cards.

6 7 8 0

How many different four-digit numbers between 6000 and 7000 can you make?

____

____

Write them in order from smallest to largest.

____

Write the largest and smallest numbers you have made in words.

____

____

10. Circle the diagram that shows 1·6 and explain your answer.

a 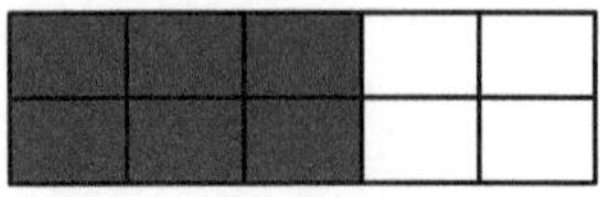

b 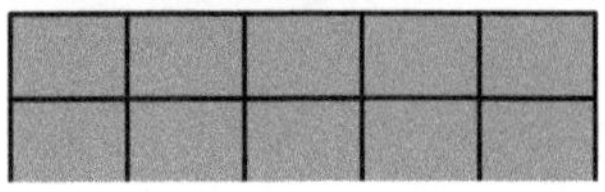

c 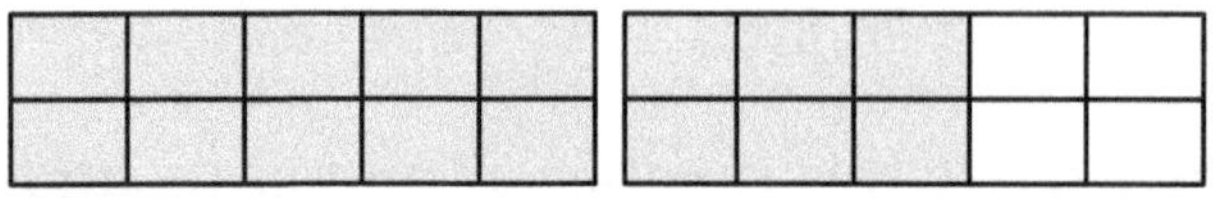

11. Write the decimal fraction each model shows.

a 

____

b  ______________

c 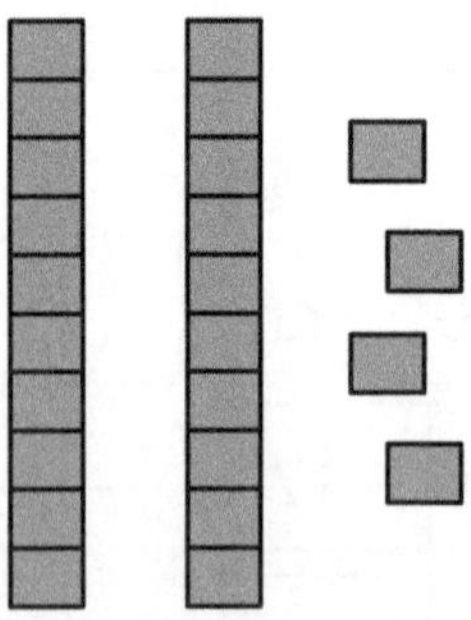 ______________

d  ______________

12. a Place the decimal fractions on the correct place on the number line:

0·6 1·2 0·9 1·5

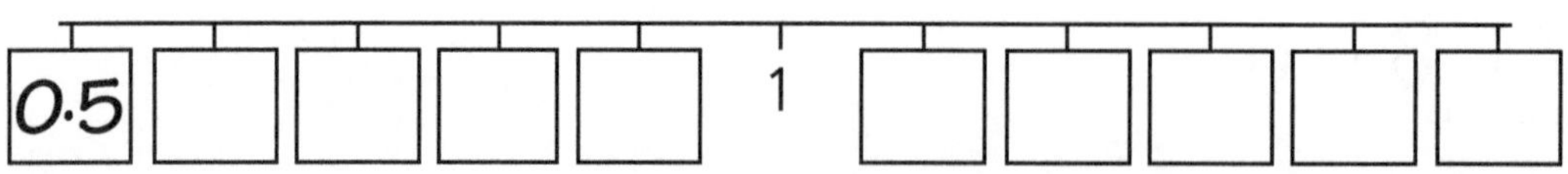

b Write the correct symbol in each box: < OR > OR =

0·9 ☐ 0·6 1·5 ☐ 1·5 0·9 ☐ 1·2

13. Using the information on the weather map, write the city name next to the correct temperature on the thermometer.

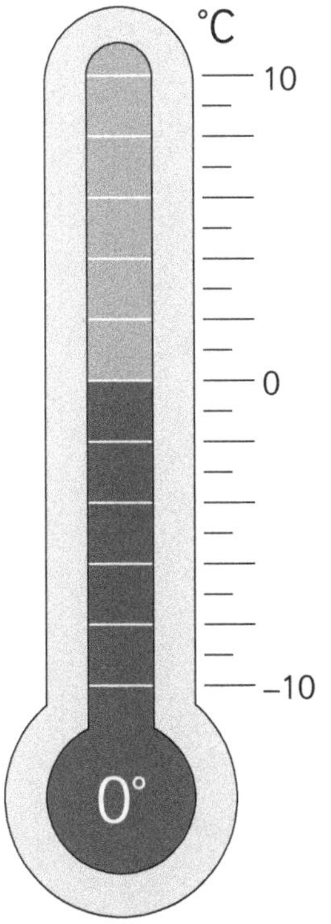

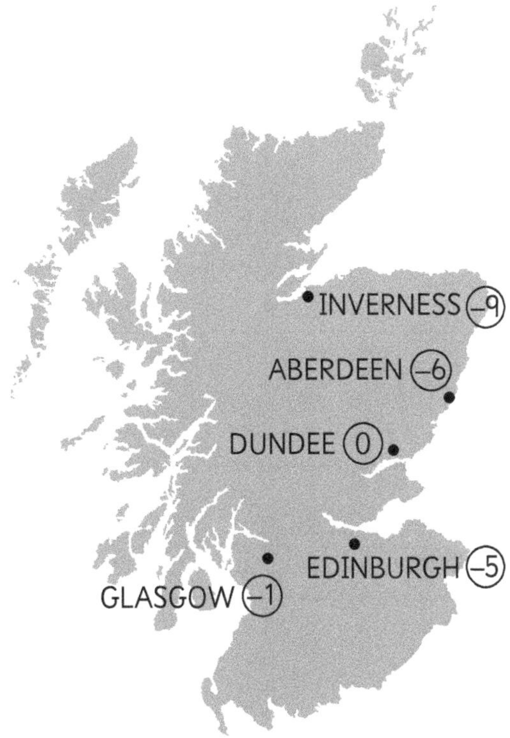

14. Using one number from the black box and one number from the white box, write and solve three different addition questions and three different subtraction questions. Explain how you worked out each answer.

225 64 78	131 29 41

15. Continue each number sequence and explain the rule.

a 8996, 8997, 8998, ________, ________, ________

The rule is ________________________________

b 3753, 3752, 3751, ________, ________, ________

The rule is ________________________________

c 3440, 3540, 3640, ________, ________, ________

The rule is ________________________________

d 3350, 3360, 3370, ________, ________, ________

The rule is ________________________________

e 1250, 2250, 3250, ________, ________, ________

The rule is ________________________________

16.

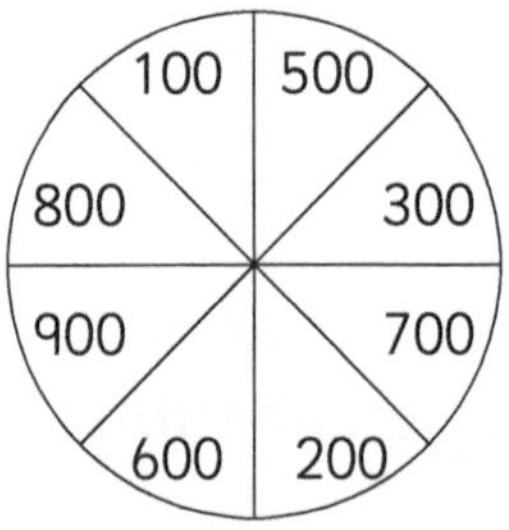

Nuria has 4 darts to throw at the dart board. If her darts land in four separate sections of the board, what is the highest number of points she can score?

____

What is the lowest?

____

What is the difference between these numbers?

____

17. Find the missing number in each of these calculations, by imagining or drawing an empty number line. Show how you can you do this in only two jumps.

a 900 – ☐ = 773 b ☐ + 289 = 600 c 1000 – 381 = ☐

Yearly progress check 2A

18. Work out the missing numbers on the number line below, then write one addition number sentence and one subtraction number sentence to match the completed number line.

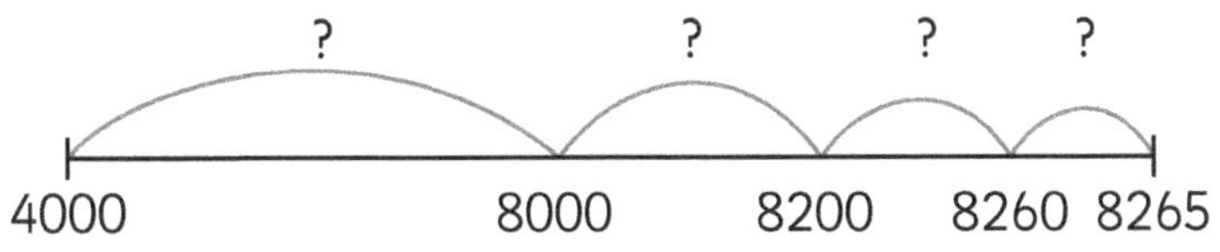

____________ ____________

Now write whether you would use counting on or counting back to solve these number sentences. Draw an empty number line beside each question to show how you solved the problem using your chosen strategy.

a [ ] + 2468 = 8000

b 9000 = 7540 + [ ]

c 3000 – 1478 = [ ]

19. Calculate the following. Explain, either with mental strategies and jottings or by drawing empty number lines, how you worked each answer out.

a 6794 – 760

b 2385 + 4400

c 8858 – 3250

Replace the ✶ with a multiple of 10 or 100 to make each number sentence true.

d 2335 + ✶ = 2935

e 3587 – ✶ = 3517

f 6745 – ✶ = 6605

20. Isla is saving up for a new bike costing £170. So far, she has saved £84. She gets £50 for her birthday. How much money does Isla still need? You may wish to draw a bar model to help you work this out.

21. Draw a Think Board with 4 sections – 'Word Problem', 'Bar Model', 'Empty Number Line' and 'Number Sentence'. Write your own Word Problem and fill in each section for 1406 – ☐ = 754

22. Write the total represented by the set of place value counters. Partition this total in three different ways.

___

___

___

___

23. Calculate the following, and show how you worked out your answers:

    a 175 + 579

    b 486 + 354

    c 848 + 484

24. Calculate the following, and show how you worked out your answers:

    a 434 + 588 + 679

    b 616 + 424 + 55

    c 788 + 233 + 923

25. Represent and solve these word problems using a method of your choice.

    a Amman is visiting his relatives for the weekend who live 144 miles away. His family stop the car for a break but still have 77 miles to go. How far have they travelled?

    b Amman and his relatives visit Stirling Castle. The gift shop sells some pencils, 209 rubbers and 362 mugs. If 687 gifts were sold, how many pencils does the gift shop sell?

26. Copy and complete the following number sentences by finding the number hidden by the stars.

    a ✶ × 2 = 12

    b 5 × ✶ = 25

    c 7 × 10 = ✶

    d 2 × ✶ = 16

    e ✶ × 8 = 80

    f 5 × ✶ = 20

    g 35 = ✶ × 7

    h 100 = 10 × ✶

27. Use your knowledge of multiplication facts to complete these triangles.

a
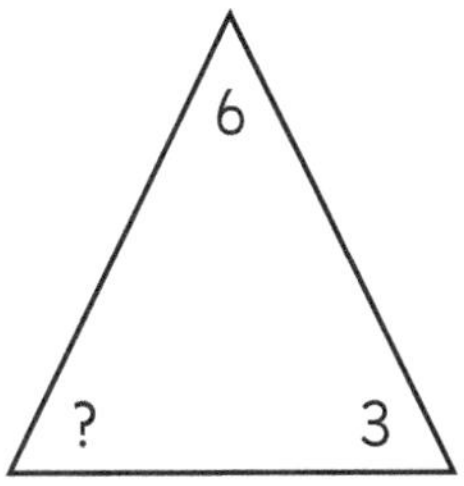

b
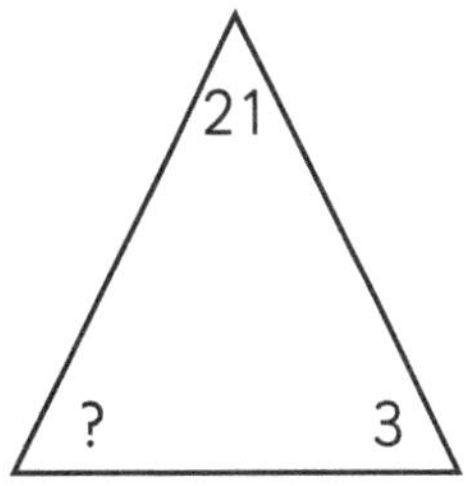

c
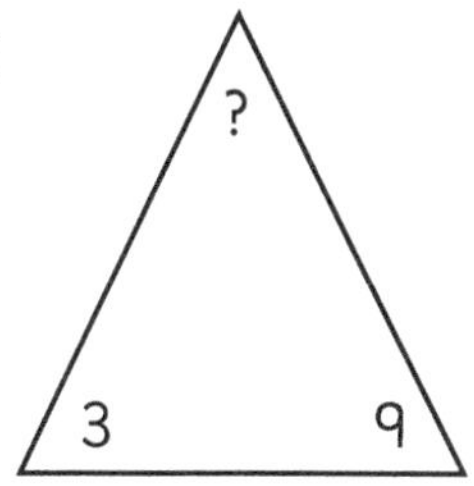

d Now write two multiplication number sentences and two division number sentences for each triangle.

28. a This is an array for 8 × 7.

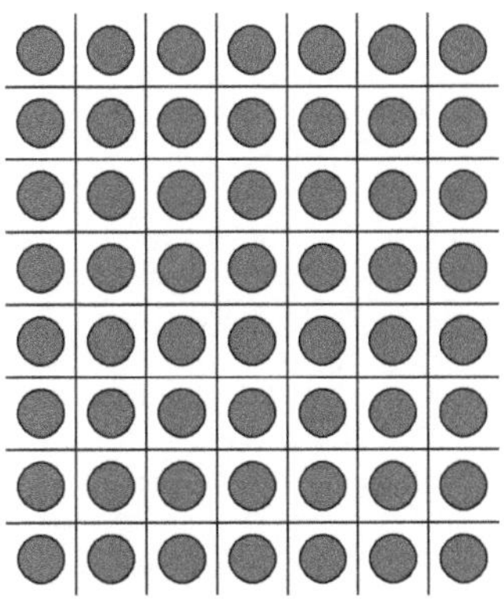

Use your pencil to divide the array so that you could use multiplication facts for 2, 3, 4 or 5 to find the answer.

b How many cans would there be in 7 rows with 12 cans in each row?

You can draw an array or a diagram to help you work out this calculation. You may also want to use bracket notation to help you record this as a number sentence.

29. a There are 100 packets of 15 felt pens in a box. Finlay says that means there are 150 felt pens altogether, Amman thinks 1500. Who is correct? Explain your thinking.

b Finlay thought of a number. He multiplied it by 10 and then by 10 again. He now has 800. What number did he start with? Explain your answer.

30. Use the multiplication square to show how the 4 times table is linked to the 2 times table.

×	1	2	3	4	5	6	7	8	9	10
1										
2										
3										
4										
5										
6										
7										
8										
9										
10										

31. Isla, Nuria and Amman each have a bag of 24 small sweets.
Amman eats 3 sweets each day.
Nuria eats 4 sweets each day.
Isla eats 6 sweets each day.

a Who will finish all of their sweets first and how many days will they last?

____

b Whose sweets will last the longest and how many days will they last?

____

32. Change the order of the factors for the following multiplication sentences. Draw an array for each fact. For example, in question **a** you should draw an array for 3 × 7 and another array for 7 × 3.

    a 3 × 7   b 9 × 2   c 4 × 6

33. Finlay and his dad have baked 32 cupcakes for the school fair. They share the cupcakes out so there are 4 cupcakes in a bag.

    a How many bags do they need? ________

    b How does you answer change if they put 8 cupcakes in each bag?

    ____

34. Are these statements true or false? If they are false, change the statement so it is correct.

    a 20 is a multiple of 4. ________________________

    b 5 is a factor of 14. ________________________

    c 2 and 5 are both factors of 10. ________________________

    d The first three multiples of 3 are 12, 15 and 18. ________________________

35.

one half	

one third	**one third**	

one quarter	**one quarter**	**one quarter**	

one fifth	**one fifth**			

one sixth	**one sixth**				

one eighth	**one eighth**	**one eighth**	**one eighth**	**one eighth**	**one eighth**	**one eighth**	

Write the correct fraction words to complete the fraction wall.

36. Using your completed fraction wall above, identify 5 sets of fractions that are equivalent. Write these below, using the = sign to show they are equivalent.

37. 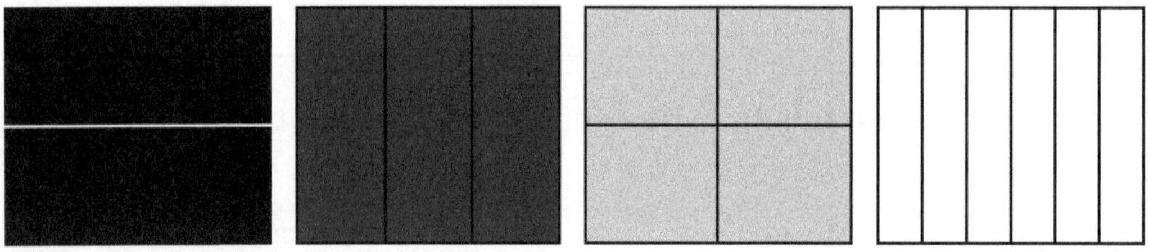

These pieces of paper have been split into halves, thirds, quarters and sixths. Write four different equivalent fractions that you can see.

38. Name these fractions. Write each fraction in its simplest form.

a 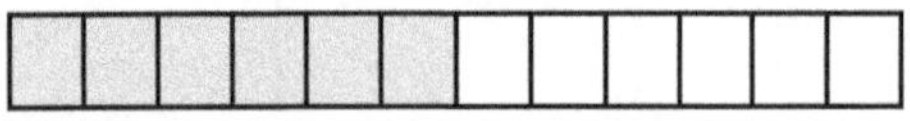

b 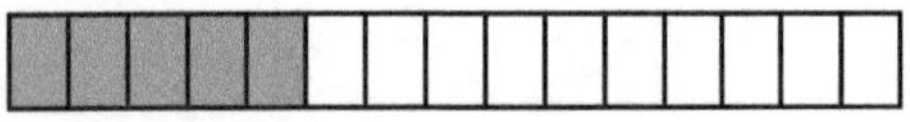

c 

d 

39. a How much pizza is there?

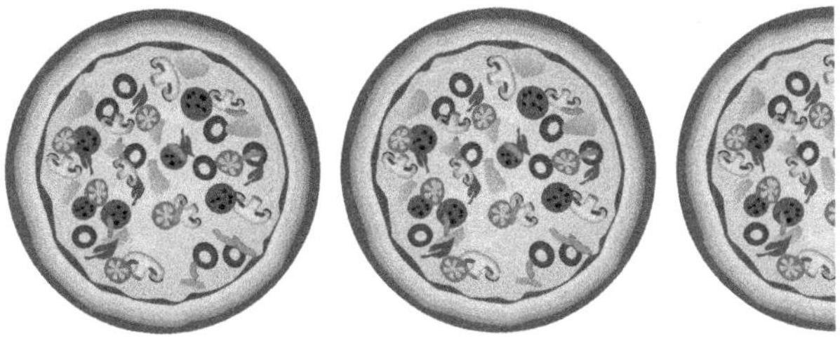

____

b How much cheesecake is there?

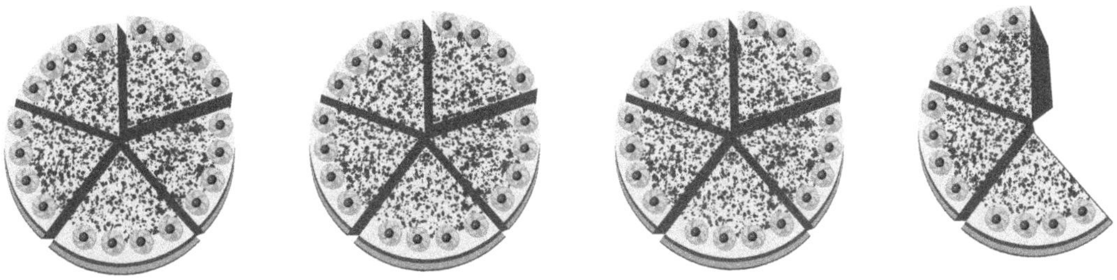

____

40. a Write the following decimal fractions in order from the smallest to the largest:

19·9 1·5 19·5 20·1 9·8

____

b Write a decimal fraction that lies in between each of the numbers in your list.

____

41. Draw a bar model to represent these word problems and solve them.

a Isla is going on a 27 km bike ride with her mum. They plan to stop for a snack when they have cycled four ninths of the way. How far is it to their snack?

b Five eighths of the children in the class are boys. How many girls are in the class if there are 32 children altogether?

Yearly progress check 2A

42. Write the following amounts using decimal notation:

a 427p ____________ b 75p ____________

c Ten pounds and two pence ____________ d 100 one pence pieces ____________

e Eight £5 notes ____________ f 14 five pence pieces ____________

43. Amman bought two concert tickets for £65 each. Unfortunately, he was ill and could not go, but he managed to sell both tickets to friends for £110 in total. Did Amman make a profit or loss and by how much?

44. Convert the times from analogue to digital or from digital to analogue. Write each answer in words using 12- and 24-hour time.

a

____

b

____

c

____

45. Complete this table, filling in the missing times.

12-hour time	24-hour time
midnight	
3:20 a.m.	
	06:30
noon	
	15:27

46. Write the times shown on each clock using 'quarter to' or 'quarter past'.

a 

____

b 

____

c 

____

d Isla is watching a tv programme that is three quarters of an hour long. It starts at 2:45 p.m. What time does it finish? Write your answer in three ways – digital, analogue and in words.

____

____

____

Yearly progress check 2A

47.

Edinburgh	**0830**	**0900**	**0930**
Haymarket	**0840**	**0910**	**0940**
Edinburgh Park	**0850**	**0920**	**0950**
Linlithgow	**0900**	**0930**	**1000**
Polmont	**0910**	**0940**	**1010**
Falkirk Grahamston	**0920**	**0950**	**1020**
Camelon	**0930**	**1000**	**1030**

Use this timetable to answer the following questions:

I arrive at Haymarket train station at 9:15 a.m. to get a train to Polmont.

a How long do I need to wait until the next train leaves? ____________

b How long does the train take to get to Polmont? ____________

c How much longer does it take the train to arrive at Camelon after it leaves Polmont? ____________

48. You will need a ruler.

Estimate the length of each of the bars below in centimetres. Write your estimate beside the bars.

Using a ruler, measure the length of the bars and write these in centimetres and in millimetres.

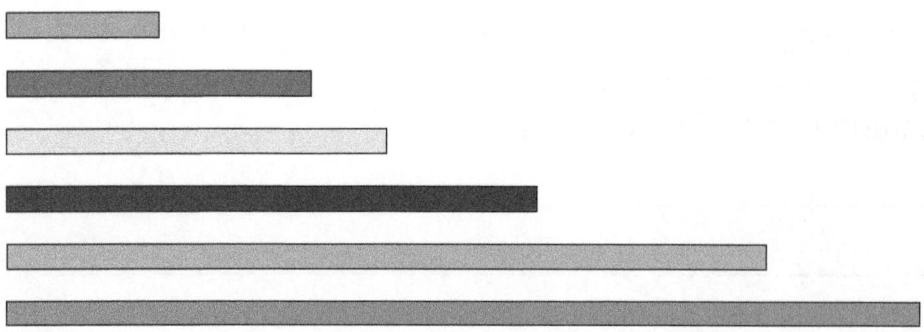

49. What is the mass being weighed on each of the scales below?

____

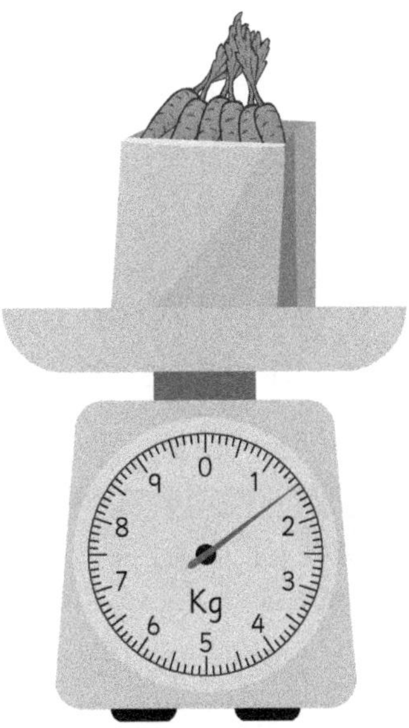

____

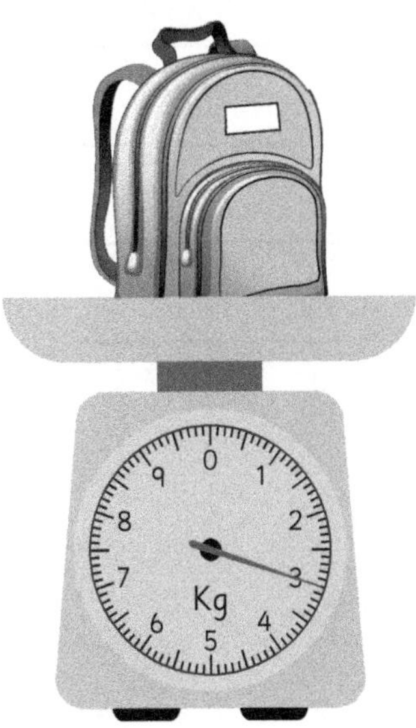

____

50.

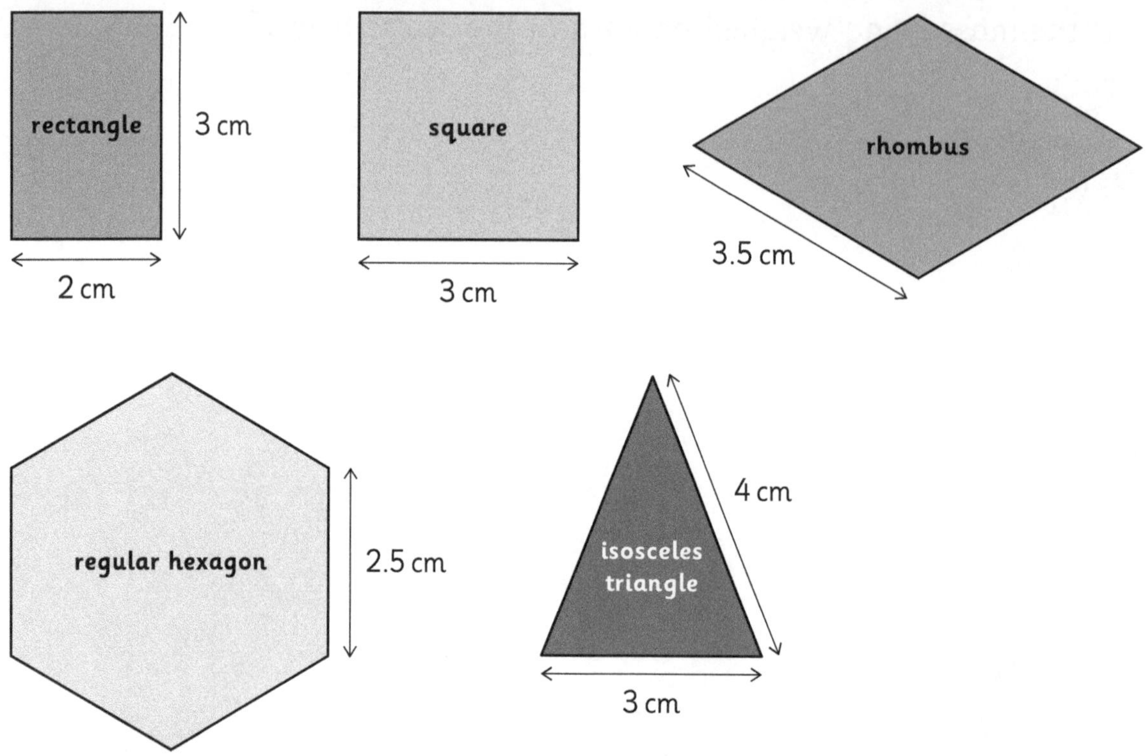

Use a ruler to measure and calculate the perimeter of each shape. Write the fewest number of measurements you will need to take for each shape and explain your thinking. Complete the table below:

Shape	Fewest number of measurements	Perimeter
rectangle		
square		
isosceles triangle		
rhombus		
regular hexagon		

Yearly progress check 2A

51. Use the dimensions to calculate the area of the shapes below in square centimetres.

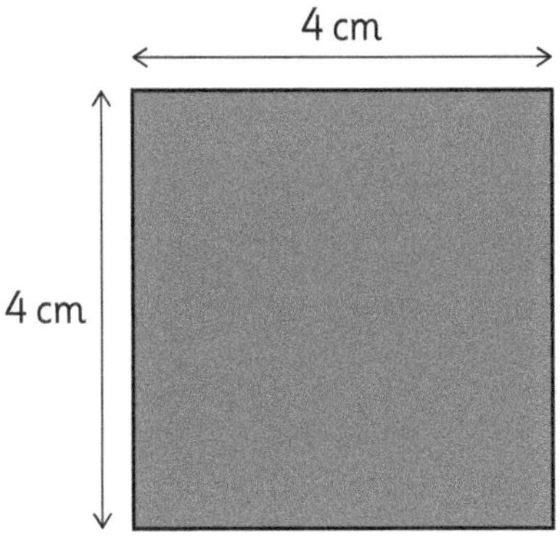

____

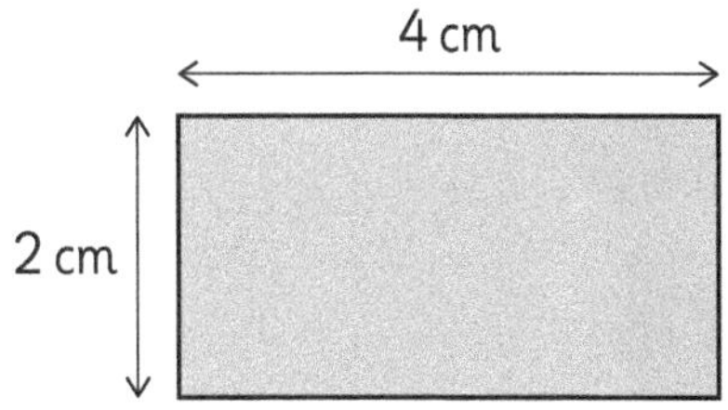

____

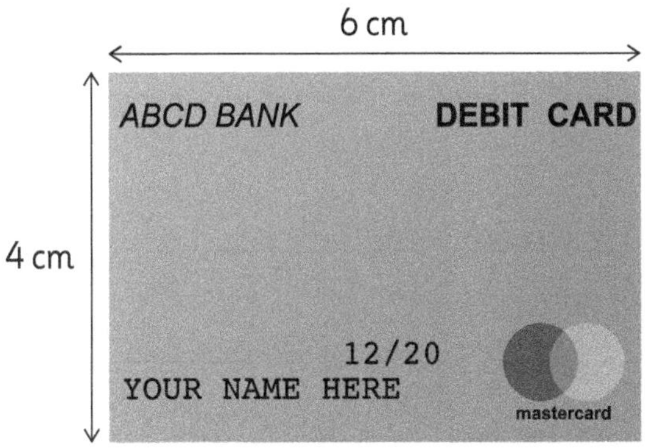

____

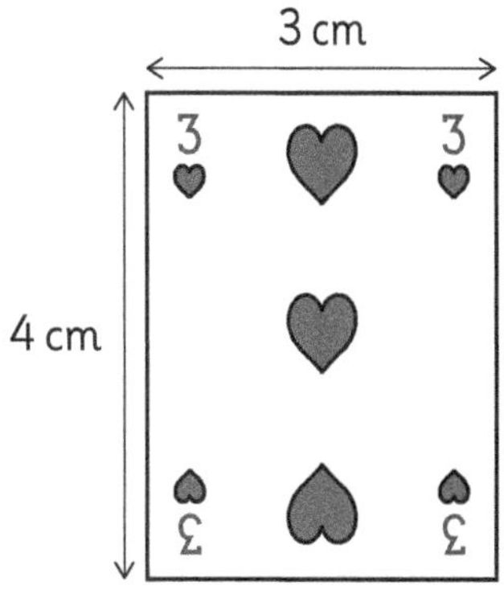

____

52. How much liquid is there in each of the containers shown here?

a

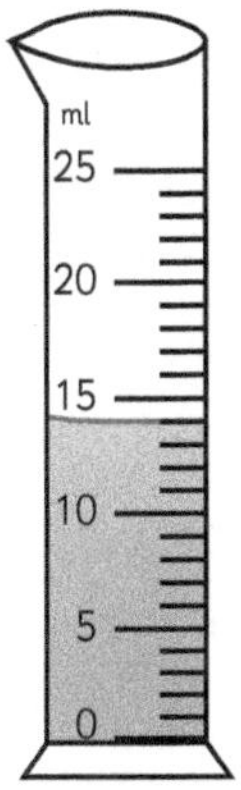

____

b

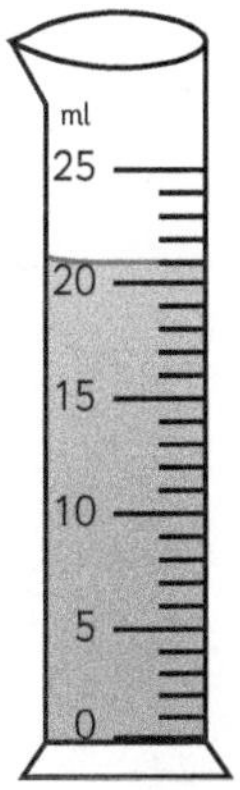

____

c

____

d

____

e

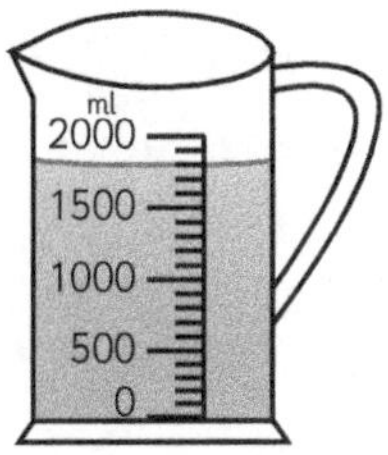

____

Yearly progress check 2A

53. Find the volume of these objects by re-creating the model and counting the cubes. Remember that some of the cubes are hidden.

a 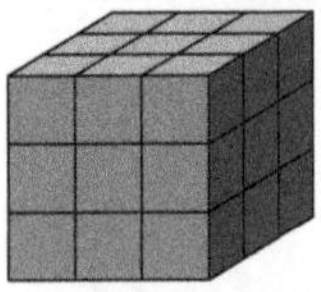 ________

b 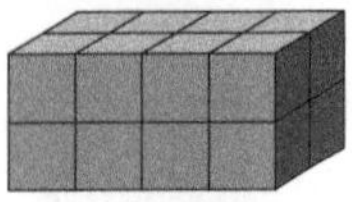 ________

c 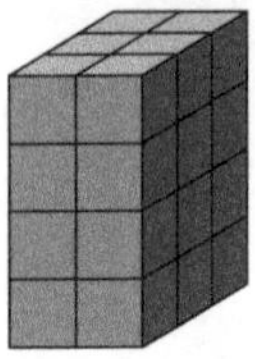 ________

d 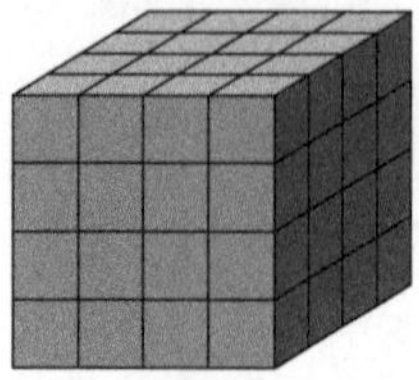 ________

e  ________

54. Complete the following patterns and sequences. Explain the rule for each pattern or sequence.

a ✶ ✶ ✶ ☾ ✶ ✶ ✶ ☐ ☐ The rule is: ________________________

b Z, X, V, T, ☐ ☐ The rule is: ________________________

c 59, 50, 41, 32, 23, ____, ____. The rule is: ________________________

d 1, 2, 4, 7, 11, 16, ____, ____. The rule is: ________________________

55. Read the word problems and write the equation that matches the problem.

a If one spider has 8 legs, how many spiders would you need for there to be 32 legs in total?

____

b I think of a number, then I divide it by 4. The answer is 6. What was my number?

____

c Finlay is at the park with his dog. All the dogs have 4 legs and he can see 28 legs in total. How many dogs are there?

____

56. You will need a protractor.

Using your protractor:

a draw a square with sides of length 5 cm.

b draw a 3 cm by 4 cm rectangle as accurately as you can.

57. 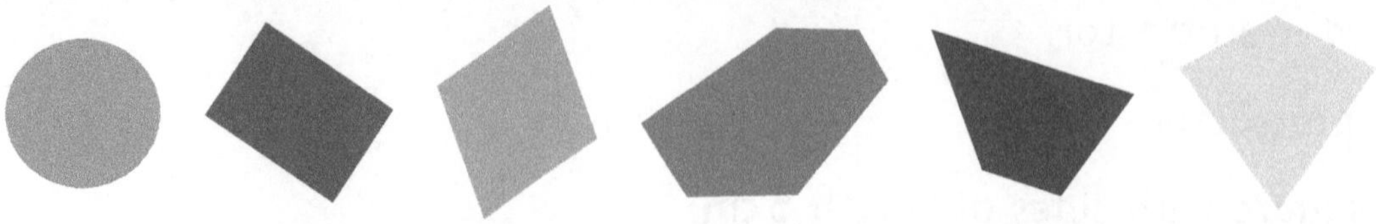

Identify the properties of the shapes shown above. Sketch the shapes that match the description into the table below.

Has straight sides	Has exactly four sides	Has pairs of parallel sides	Has square vertices	Other

58. Isometric dotty paper is needed for this question.

Draw the top view, one side view and one end view for the object below on isometric dotty paper.

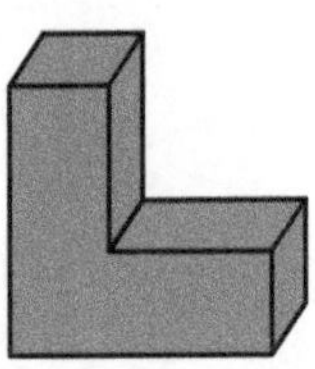

59. 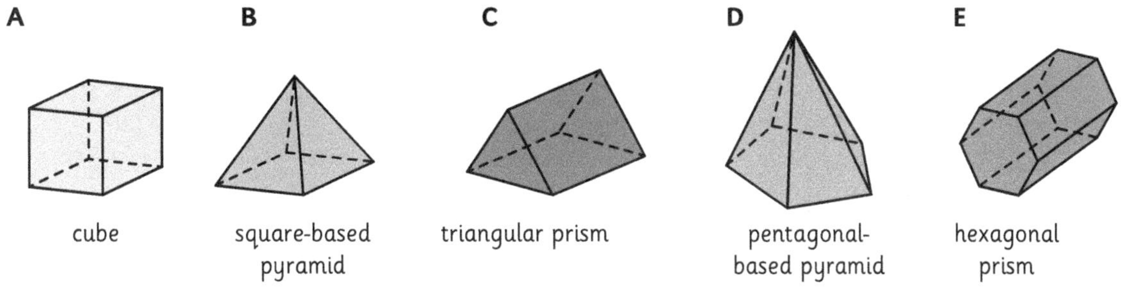

Complete the Venn diagram by writing the letter of the 3D objects A to E in the correct place.

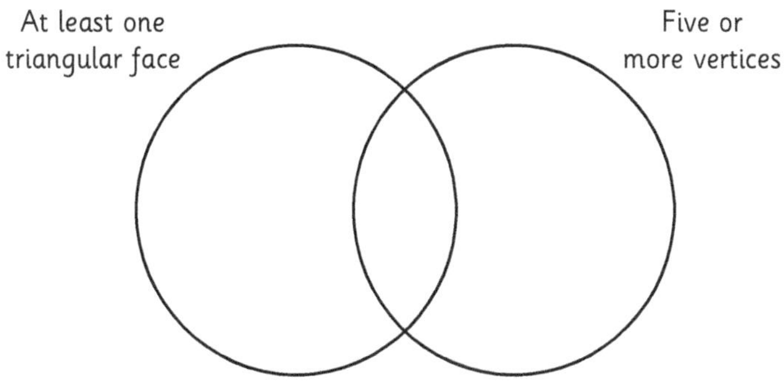

60. Can you spot the acute, right and obtuse angles? Write the letters next to the correct type of angles in the table below.

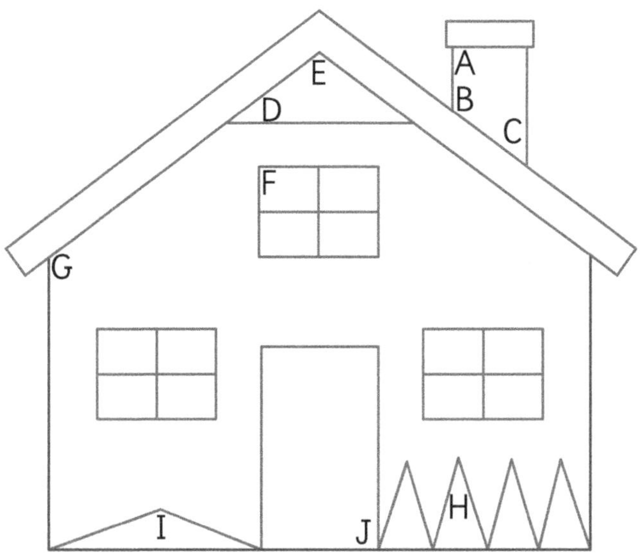

Acute	
Right	
Obtuse	

Yearly progress check 2A

61. Using the squared paper below, sketch the first initial of Finlay's name. Choose a starting point and mark this on the letter you have drawn. Write directions from the start point so that a friend could draw the same letter exactly like yours. Use directions such as North 2, East 5, etc.

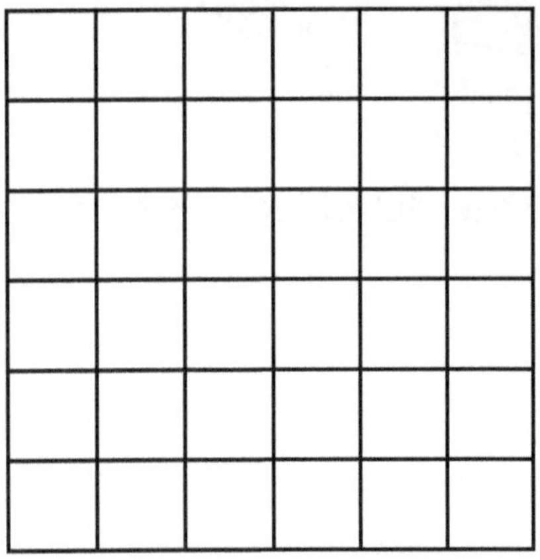

62. a A square has been drawn on the grid. Write the coordinates of its vertices.

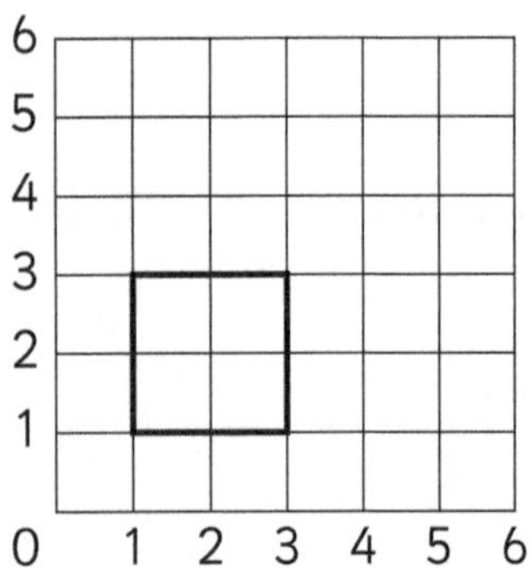

b A rectangle has the vertices (6, 6), (1, 6), (6, 4) and (1, 4). Plot these points on the coordinate grid and draw the rectangle.

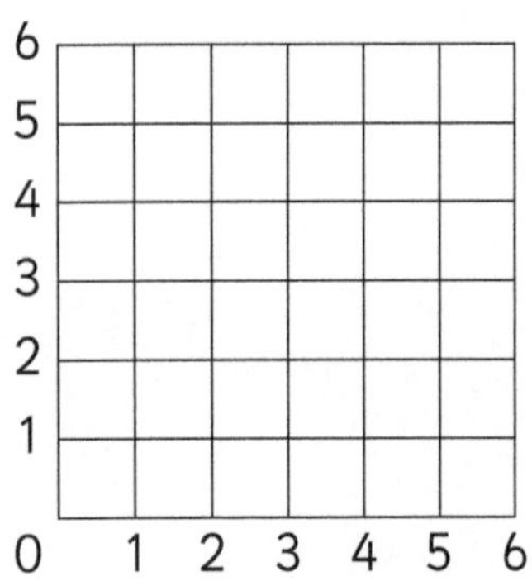

Yearly progress check 2A

63. Draw the lines of symmetry onto these shapes, if there are any. Complete the table below to sort the shapes.

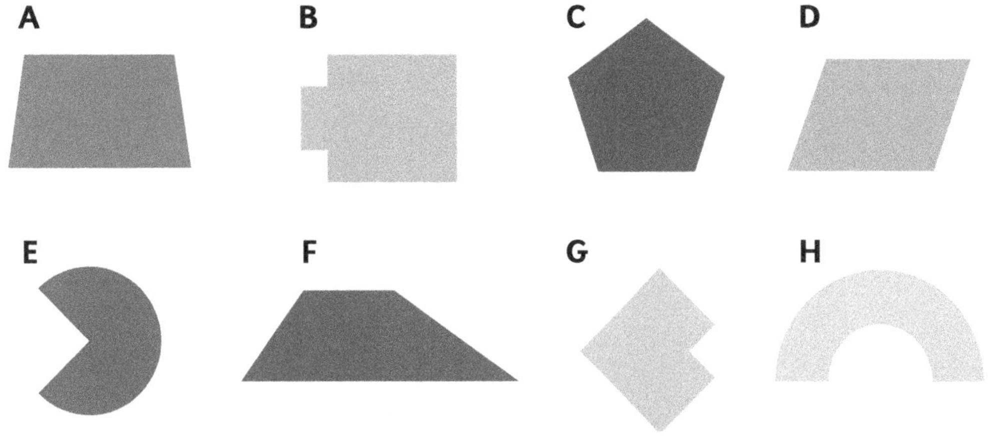

Horizontal line of symmetry	Vertical line of symmetry	No line of symmetry

64. Draw your own design for a flag. The flag must be symmetrical at the line of symmetry and include at least four different colours.

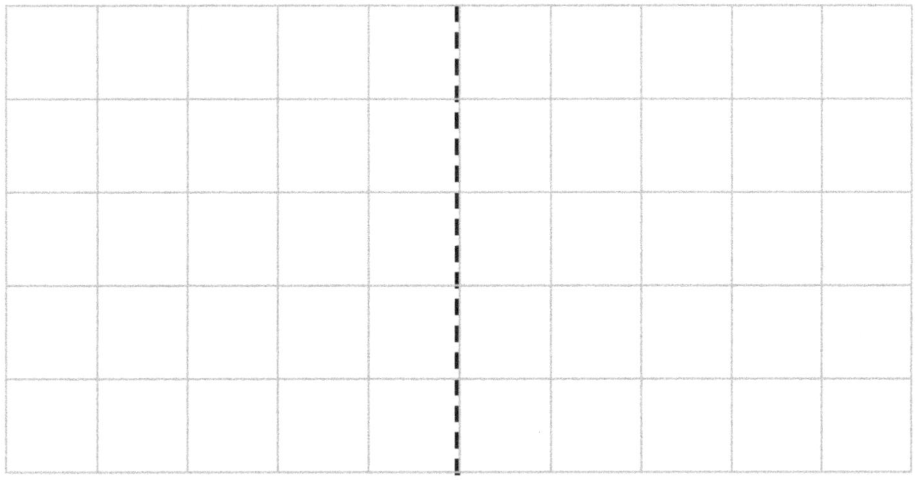

65. A protractor is needed for this section.

Use a protractor to measure each angle. Write the size of each angle you measure.

a
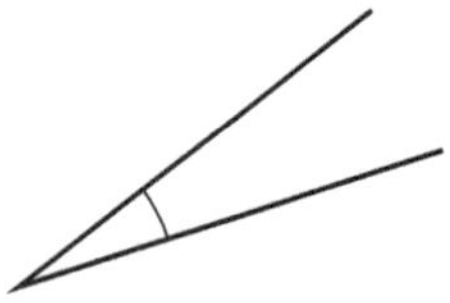
____

b
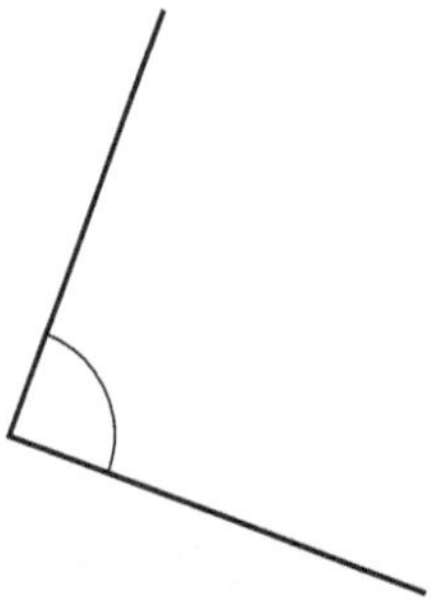
____

c
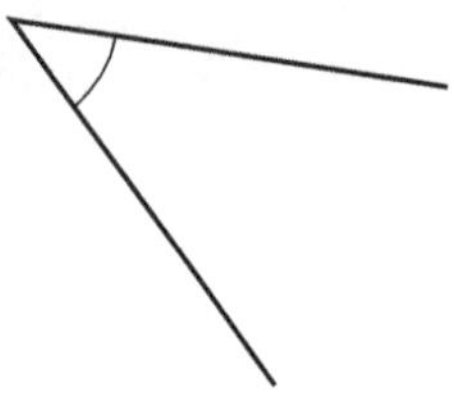
____

d
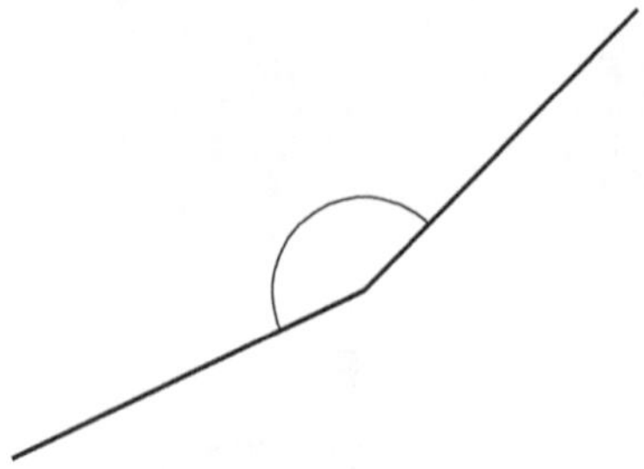
____

e

____

66. A ruler is needed for this section.

The scale on this map is 1 cm for every 40 km. For each question below use a ruler to measure the distance to the nearest whole centimetre and then convert to kilometres.

a About how far is Inverness from Aberdeen?

____

b Perth is about 80 km from which town/city?

____

c Which towns are the furthest apart? About how far is this? __________

d Which towns are nearest each other? About how far is this? __________

67. The picture graph shows how many mobile phones a company sold over five days.

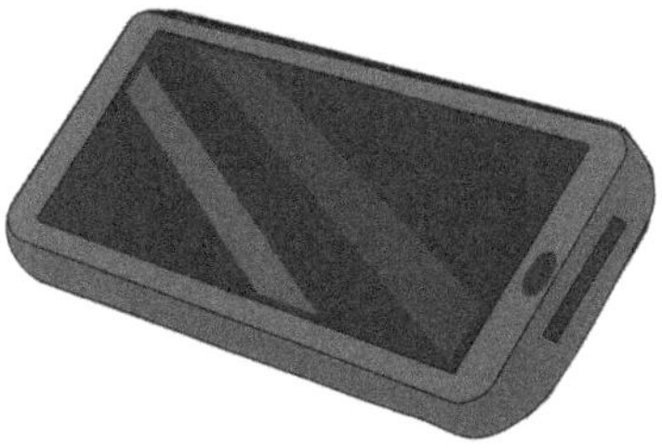

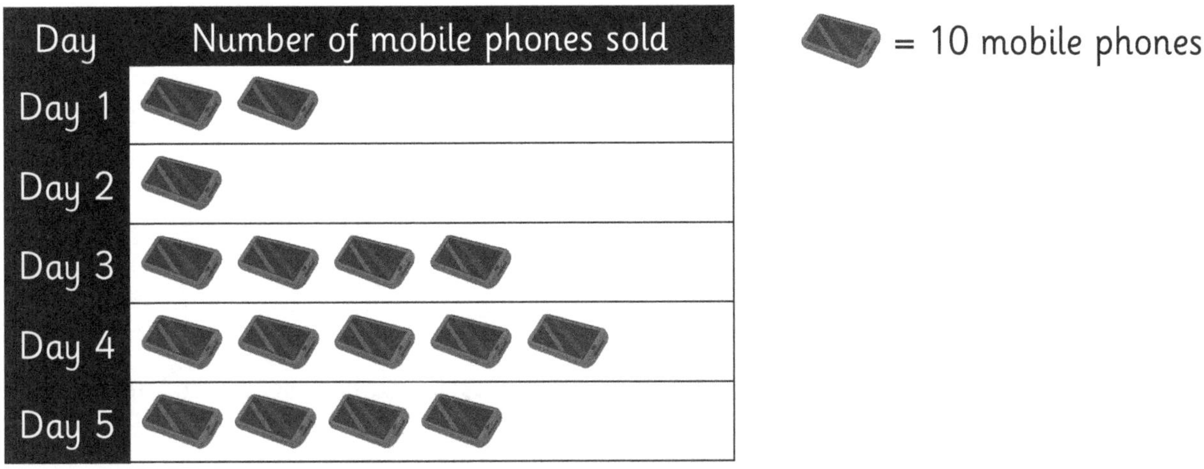

a How many phones were sold in total over the five days? ____________

b How many more phones were sold on Day 3 than on Day 1? ____________

c On which day were the most phones sold? ____________

d Were more phones sold on Days 1 and 2 or on Days 4 and 5? ____________

68. The line graph shows the number of laptops that a company sold over six months.

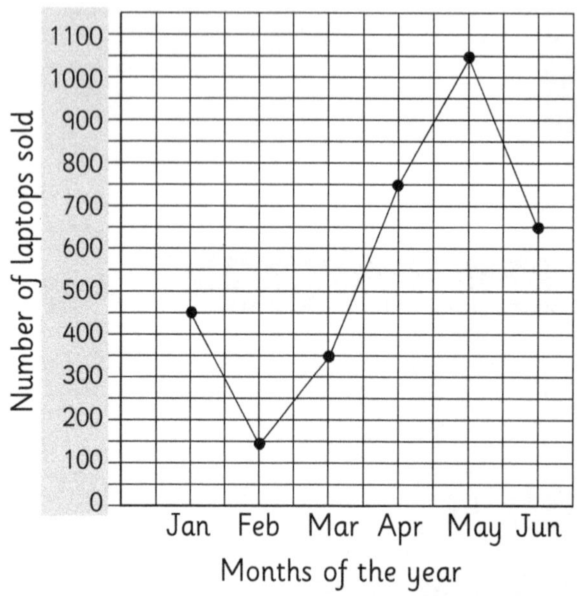

a In which months did the number of laptops sold decrease from the previous month? ___________

b How many fewer laptops were sold in February than in January?

____

c How many more laptops were sold were sold in the three months of April, May and June than in the period January, February and March?

____

Yearly progress check 2A

69. Complete the 'frequency' column in the table. Display the data in the table as a bar graph.

Snack	Tally	Frequency
Crisps	𝍸 𝍸 𝍸 𝍸 𝍸 𝍸 𝍸 |	
Fruit	𝍸 𝍸 𝍸 𝍸 𝍸 𝍸	
Sweets	𝍸 𝍸 ||||	
Biscuits	𝍸 𝍸 ||	
Vegetable sticks	𝍸 𝍸 𝍸 |||	

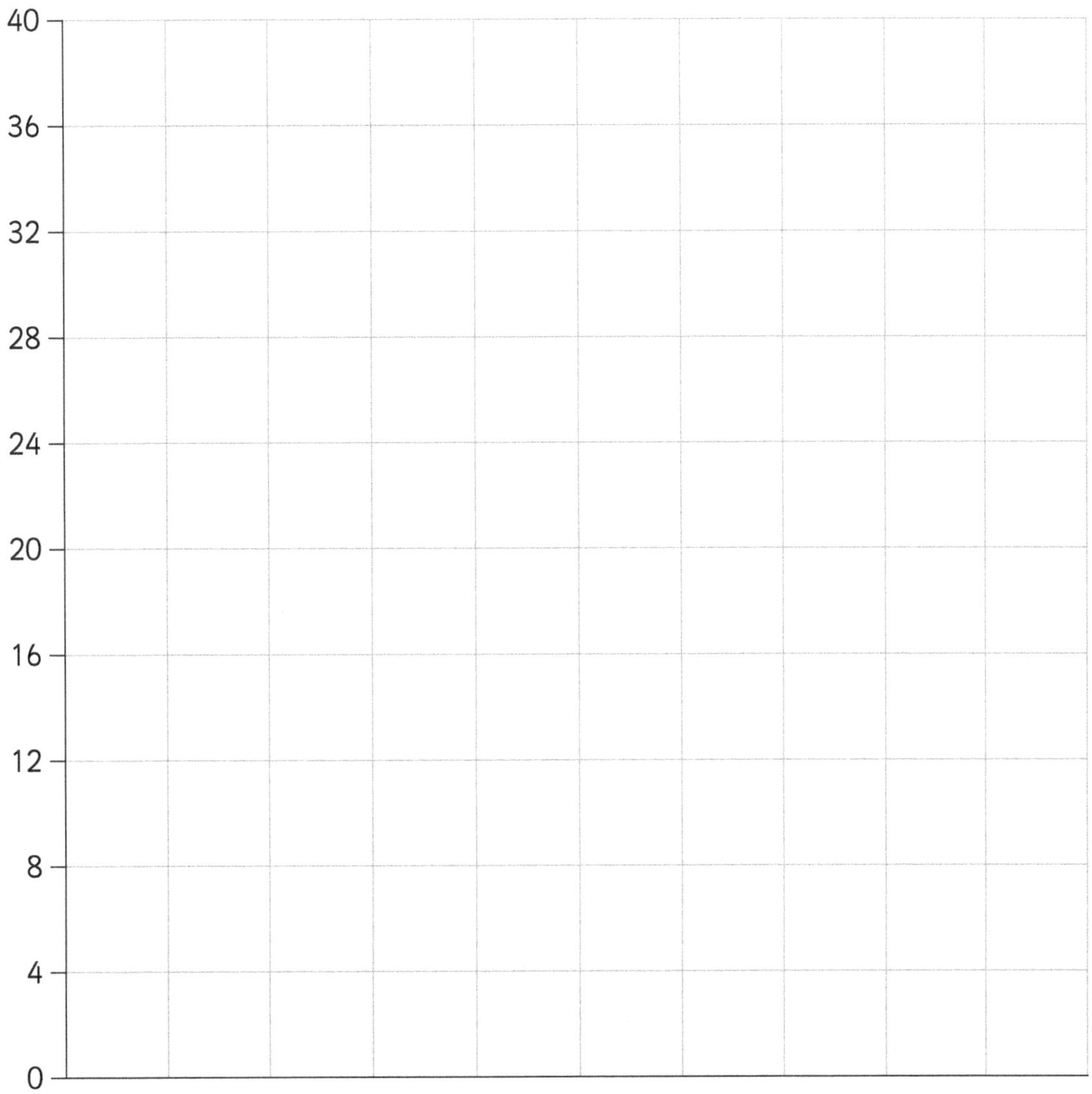

Yearly progress check 2A

70. This pie chart shows the favourite sports of 48 children at an after-school club.

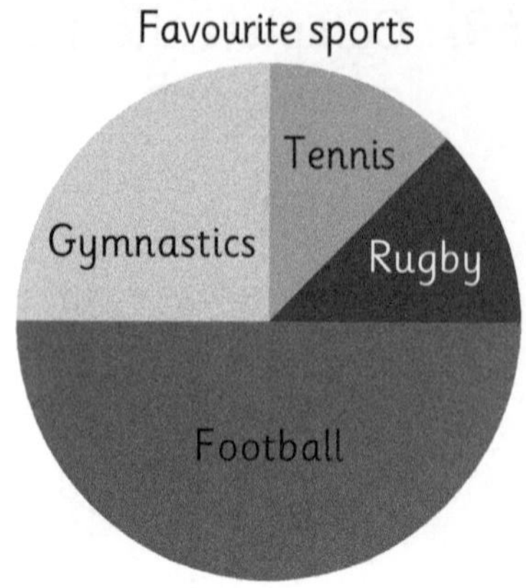

a How many children chose gymnastics? How many children chose rugby? How did you work this out?

b How many fewer children like rugby than gymnastics? How did you work this out?

71. Using the probability scale, write examples about each probability that are true for you. For example: Impossible – I will visit Mars today.

Impossible **Unlikely** **Even chance** **Likely** **Certain**

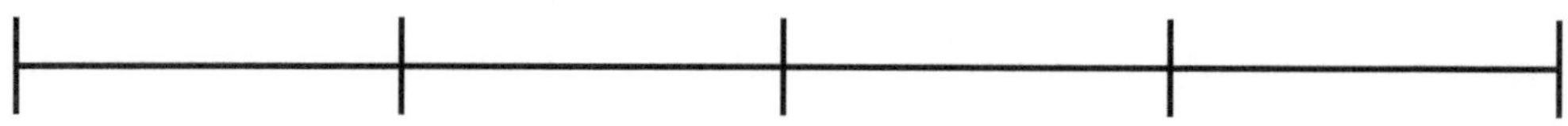

0 **0·25** **0·5** **0·75** **1**

0% **25%** **50%** **75%** **100%**

Topic–benchmarks/ Es & Os	Question	Answer	Marking guidance
Estimation and rounding MNU 2-01a	1	Three hundred and fifty-six 360 400	**On track** • Writes the number 356 accurately in words. Although accurate spelling should be encouraged, reasonable and recognisable attempts (e.g. 'hundrid') should be accepted. • Understands the conventions of rounding, rounding up when the ones digit is 5 or more. **Further learning required** • May not understand the importance of the word 'and' so thinks it can be missed out. • May use the incorrect word(s) to write the number; for example, reads '56' as sixty-five. • May use incorrect rounding strategies, e.g. focusing on the incorrect digit for rounding or rounding to the nearest hundred instead of the nearest ten.
MNU 2-01a	2	Finlay: 647 is closer to 600 than 700 and so must be rounded down to 600	**On track** • Understands the conventions of rounding and can justify why Finlay is correct or why Amman cannot be correct. **Further learning required** • May confuse when to round up or round down, e.g. uses the '7' ones to round up when rounding to the nearest 100.
MNU 2-01a	3	Numbers 0.7, 1.0, 1.4 and 1.6 should be positioned accurately.	**On track** • Decimal fractions are positioned accurately on the number line. • Understands the conventions of rounding, rounding up when the tenths digit is 5 or more. **Further learning required** • May focus solely on the tenths digit and so think that 0.7 should come last. • May be unclear about when to round up or round down and/or disregards the digit after the decimal point.
MNU 2-01a	4	Stirling: 5°C Aberdeen: 3°C Perth: 4°C Edinburgh: 5°C Oban: 3°C Stirling and Edinburgh (5°C) Aberdeen and Oban (3°C)	**On track** • Understands the conventions of rounding, rounding up when the tenths digit is 5 or more. **Further learning required** • May be unclear about when to round up or round down and/or disregards the digit after the decimal point.

Topic–benchmarks/ Es & Os	Question	Answer	Marking guidance
MNU 2-01a MNU 2-03a	5	20 + 90 = 110 21 + 87 = 108 50 + 70 = 120 45 + 67 = 112 30 + 80 = 110 34 + 75 = 109	**On track** • Rounds correctly in order to estimate the totals. • Adds the two numbers accurately using mental strategies. **Further learning required** • May use incorrect rounding strategies, e.g. focusing on the incorrect digit for rounding. • May add the two numbers inaccurately or use an inefficient strategy.
Number – order and place value MNU 2-02a	6	Smallest number: 1069 One thousand and sixty-nine Isla is correct; whole numbers do not usually start with a zero.	**On track** • Understands the purpose of the 'zero' number card in creating four-digit numbers. • Writes number words accurately. Although accurate spelling should be encouraged, reasonable and recognisable attempts (e.g. hundrid) should be accepted. **Further learning required** • May not understand zero as a place holder and/or attach a value to it and so thinks 0169 is the smallest 4-digit number. • May write four-digit numbers incorrectly, e.g. one hundred and sixty-nine instead of one thousand and sixty-nine.
MNU 2-02a	7	a 3443 Three thousand, four hundred and forty-three b Isla's number is 4027.	**On track** • Understands the relationship between the base 10 blocks and the number they represent. • Writes four-digit number words accurately. Although accurate spelling should be encouraged, reasonable and recognisable attempts (e.g. hundrid) should be accepted. **Further learning required** • May be confused where to position the place holder '0'. • May write four-digit numbers incorrectly, e.g. by confusing thousands for tens of thousands or omitting the 'and'.
MNU 2-02a	8	200 and 150 and 17 Any three numbers that total 367	**On track** • Flexibly partitions 367. **Further learning required** • Struggles to partition 367 into three parts. • Unable to 'see' 367 as anything other than 300 + 60 + 7 or 3 hundreds, 6 tens and 7 ones.

Topic–benchmarks/ Es & Os	Question	Answer	Marking guidance
MNU 2-02a	9	Six four-digit numbers can be made: 6078, 6087, 6708, 6780, 6807, 6870. Six thousand and seventy-eight Six thousand, eight hundred and seventy	**On track** • All six four-digit numbers are made, written and ordered accurately. **Further learning required** • May be confused about where to position the place holder '0' and so construct numbers such as 0678. • May write four-digit numbers incorrectly, e.g. by confusing the value of thousands. • May be unable to write all six numbers.
MNU 2-02a	10	The diagram showing 1.6 is circled and a satisfactory explanation given.	**On track** • Identifies the correct decimal fraction diagram and can justify their choice. **Further learning required** • Is unable to identify 1.6 as one whole and six tenths.
MNU 2-02a	11	a 1.3 b 1.5 c 2.4 d 1.1	**On track** • Identifies and writes the decimal fractions accurately. **Further learning required** • May not understand the concept of one whole being equal to 10 tenths. • May think they can disregard the decimal point or zeros and write, for example, 1003 or 13.
MNU 2-03b	12	a 0.6, 0.9, 1.2 and 1.5 placed correctly on the number line. b 0.9 > 0.6 1.5 = 1.5 0.9 < 1.2	**On track** • Decimal fractions are accurately placed on the number line. • <, > and = are used to complete the expressions correctly. **Further learning required** • May be confused about where to place the decimal fractions on the number line. This may be because: - they do not understand how to 'read' the number line - they do not understand that there are numbers that lie between consecutive whole numbers that are fractions. • May mix up the < and > symbols.
MNU 2-04a	13	Place names recorded alongside the thermometer in the following positions: Dundee: 0°C Glasgow: −1°C Edinburgh: −5°C Aberdeen: −6°C Inverness: −9°C	**On track** • Records the cities' temperatures accurately on the thermometer. **Further learning required** • May be confused about where to write the names of the cities on the thermometer, possibly mixing up positive and negative numbers.

Topic–benchmarks/ Es & Os	Question	Answer	Marking guidance
Number – addition and subtraction MNU 2-02a MNU 2-03a	14	Addition possibilities 225 + 131 = 356 225 + 29 = 254 225 + 41 = 266 64 + 131 = 195 64 + 29 = 93 64 + 41 = 105 78 + 131 = 209 78 + 29 = 107 78 + 41 =119 Subtraction possibilities 225 – 131 = 94 225 – 29 = 196 225 – 41 = 184 64 – 29 = 35 64 – 41 = 23 131 – 64 = 67 131 – 78 = 53 78 – 29 = 49 78 – 41 = 37	**On track** • Calculates accurately. Attention should also be paid to the numbers that are chosen, e.g. only choosing numbers that do not involve bridging a ten and/or hundred. • May use strategies such as compensation (mentally), jumps on an empty number line, counting on or informal jottings. • May also choose different strategies, depending on the calculation. • Is able to clearly explain how answers have been worked out. **Further learning required** • May calculate inaccurately; possibly due to miswriting the order of the numbers in the calculation, the strategy chosen or choosing an appropriate strategy but using it ineffectively e.g. jumping back in 10s when subtracting 131. • May choose numbers that do not require bridging of 10s or 100s or smaller numbers. This may demonstrate a lack of confidence in working with larger numbers or using bridging strategies.
MNU 2-02a	15	a 8996, 8997, 8998, **8999, 9000, 9001** The rule is: **Add 1/increasing by 1** b 3753, 3752, 3751, **3750, 3749, 3748** The rule is: **subtract 1/ decreasing by 1** c 3440, 3540, 3640, **3740, 3840, 3940** The rule is: **Add 100/increasing by 100** d 3350, 3360, 3370, **3380, 3390, 3400** The rule is: **Add 10/ increasing by 10** e 1250, 2250, 3250, **4250, 5250, 6250** The rule is: **Add 1000/ increasing by 1000**	**On track** • Is able to continue each number sequence accurately. • Can accurately describe the sequence using appropriate vocabulary, e.g. increasing by 10, subtract100 each time. **Further learning required** • May be unable to accurately complete sequences that bridge a 10, 100 and/or 1000. • May be able to complete sequences that require them to count forwards and backwards in ones but find counting in 10s, 100s or 1000s more challenging. • May find counting forwards easier than counting backwards.

Topic–benchmarks/ Es & Os	Question	Answer	Marking guidance
MNU 2-02a MNU 2-03a	16	Highest possible score: 600 + 700 + 800 + 900 = 3000 Lowest possible score: 100 + 200 + 300 + 500 = 1100 Difference: 1900	**On track** • Is aware of how to construct the highest and lowest scores possible from the numbers given on the dart board. • Is able to manipulate the four numbers in order to carry out calculations accurately. • May look for doubles/near doubles or numbers that make a 'ten' when adding. • Understands that the numbers can be added in any order without affecting the total and chooses the most efficient method, e.g. 200 + 300 then adding 500, then 100. • May use compensation or count on to find the difference between the highest and lowest scores. **Further learning required** • May be unable to choose the four numbers that would give the highest/lowest score. • May calculate inaccurately, possibly due to the strategy chosen or using it ineffectively, e.g., making mistakes when bridging a ten, or not using numbers that are easier to add together.
MNU 2-02a MNU 2-03a	17	a 900 – **127** = 773 b **311** + 289 = 600 c 1000 – 381 = **619**	**On track** • Can count on or count back on a number line, either imagined, supported by jottings, or drawn, to complete each calculation in only two jumps. • Can confidently partition numbers to make calculations more manageable. • Will be able to bridge tens and hundreds without difficulty. **Further learning required** • May be unable to complete the calculations in only two jumps. • May struggle when bridging tens or hundreds. • May need to draw a number line but does so inaccurately (e.g. by writing the larger number on the left when subtracting) or is unable to represent partitioning and calculations on the number line.

Topic–benchmarks/ Es & Os	Question	Answer	Marking guidance
MNU 2-03a	18	4000 + 4265 = 8265 8265 – 4265 = 4000 a **5532** + 2468 = 8000 Will most likely count on from 2468 up to 8000. May subtract 2468 by counting back from 8000. May count back to 2468 from 8000. b 9000 = 7540 + **1460** Will most likely count on from 7540 to 9000. May subtract 7540 by counting back from 9000. May count back to 7540 from 9000. c 3000 – 1478 = **1522** Will most likely subtract 1478 by counting back from 3000. May count back to 1478 from 3000. May count on from 1478 to 3000. Number lines drawn that accurately show the method chosen.	**On track** • Can accurately write one addition and one subtraction number sentence to match the number line shown. • Can draw empty number lines to support addition and subtraction calculations. • Understands the inverse relationship between addition and subtraction so can choose between counting on and counting back to solve missing addend and missing difference problems. **Further learning required** • May be unable to write two number sentences to match the completed number line. • May be confused between adding/subtracting 10s/100s/1000s. • May be able to work out the answer when the whole is unknown (question c) but have difficulty with part unknown problems (questions a and b). For example, adds 2468 and 8000 in question a. • May use the number line incorrectly.

Topic–benchmarks/ Es & Os	Question	Answer	Marking guidance
MNU 2-03a	19	a 6794 − 760 = **6034** b 2385 + 4400 = **6785** c 8858 − 3250 = **5608** d 2335 + **600** = 2935 e 3587 − **70** = 3517 f 6745 - **140** = 6605	**On track** • Can accurately calculate and use written jottings or an empty number line to show how each calculation was worked out. • Can partition numbers flexibly and use counting on/back strategies to calculate accurately. **Further learning required** • May rely on vertical algorithms to solve simple missing sum and missing difference problems (questions a,b,c). • May be unable to choose an appropriate method to solve missing addend and missing subtrahend problems (questions d,e,f). • May add/subtract small numbers multiple times instead of a larger number. • May calculate inaccurately. • May be unable to partition effectively to help support their calculations. • May be confused with place value, e.g. use a '7' instead of '70'.
MNU 2-03a	20	£84 + £50 = £134 £170 − £134 = £36 Isla still needs £36.	**On track** • Can interpret the problem and represent it as a bar model. • Can flexibly partition the numbers in the calculation and solve it using mental strategies, supported by an empty number line or informal jottings. **Further learning required** • May struggle to understand and/or visualise the word problem to then represent this in a bar model/diagram. • May leave out parts of the calculation, e.g. only calculates how much money Isla currently has, not how much she still needs.

Topic–benchmarks/ Es & Os	Question	Answer	Marking guidance
MNU 2-03a	21	1406 – **?** = 754 The pupil should draw and complete a Think Board. As this is a Change Unknown type of question, the word problem the pupil has written should reflect this and not include the unknown number (652).	**On track** • Can write a Change Unknown word problem where the reader is given the start number and the result. • Can confidently complete all parts of the Think Board. These should relate clearly to the word problem the pupil has written. • Can flexibly partition numbers and use an effective strategy (e.g. count on or count back) to calculate the answer. • The answer of 652 is displayed in the Bar Model, Empty Number Line and Number Sentence parts of the Think Board. **Further learning required** • May struggle to write a Word Problem to match the given calculation. May include the numbers 1406 and 754 but not be able to write a Change Unknown type of problem and may revert to a Result Unknown (1406 – 754 = ?) which is incorrect. • May only be able to complete some parts of the Think Board e.g. the Number Sentence section is completed first and others are left incomplete. • May use ineffective strategies to complete their calculation.
MNU 2-02a MNU 2-03a	22	547 partitioned in three different ways, e.g. 500 + 40 + 7 (standard place value partitioning) 400 + 140 + 7; 300 + 240 + 7 (non-standard place value partitioning)	**On track** • Can use standard and flexible place value partitioning to partition 547 in three different ways, i.e. understands that while 547 is 500 + 40 + 7 it can also be 500 + 47; 540 + 7; 400 + 140 + 7; 300 + 240 + 7 etc. **Further learning required** • May only be able to use standard partitioning e.g. 547 = 500 + 40 + 7. • May demonstrate inaccurate partitioning e.g. 300 + 200 + 7.

Topic–benchmarks/ Es & Os	Question	Answer	Marking guidance
MNU 2-03a	23	a 175 + 579 = **754** b 486 + 354 = **840** c 848 + 484 = **1332**	**On track** • Can calculate accurately and explain their method. • Is confident using a variety of mental strategies/informal jottings that demonstrate flexible partitioning (e.g. jump and split strategies), bridging 10s and 100s and/or compensation. **Further learning required** • May calculate inaccurately or be unable to count on and back in multiples. • May rely on jumps of 100, 10 and 1 only when using an empty number line. • May default to the same one strategy for each calculation, e.g. vertical algorithms. • May use strategies such as compensation incorrectly e.g. may choose to add 580 in the first example but forget to take the '1' away or add it instead.
MNU 2-03a	24	a 434 + 588 + 679 = **1701** b 616 + 424 + 55 = **1095** c 788 + 233 + 923 = **1944**	**On track** • Can calculate accurately and explain their method. • Can choose an effective strategy for calculation, based on the numbers involved. • Can flexibly partition the numbers in order to calculate. **Further learning required** • May calculate inaccurately and/or be unable to count on and back in multiples. • May default to the same strategy for every calculation, e.g. vertical algorithms. • May be unable to count in multiples and rely on jumps of 100, 10 and 1 only when using an empty number line. • May add the numbers in the order they appear in the calculation, e.g. adding 434 to 588 then adding 679, instead of perhaps spotting that 400 + 600 = 1000. • May use strategies such as compensation incorrectly, e.g. may choose to add 680 in the first example but forget to take the '1' away or adds it instead.

Topic–benchmarks/ Es & Os	Question	Answer	Marking guidance
MNU 2-03a	25	a 67 miles b 116 pencils	**On track** • Can confidently demonstrate how each problem was solved by explaining mental strategies or by showing informal jottings and diagrams. • Can choose an effective strategy for calculation, dependant on the type of calculation and the numbers involved. • Can flexibly partition the numbers in order to calculate. **Further learning required** • May calculate inaccurately or choose an ineffective or inefficient strategy. • May be reliant on using the same strategy for every calculation. • May use strategies such as compensation incorrectly, e.g. may choose to add 210 instead of 209 but forget to take the '1' away or adds it instead.
Number – multiplication and division MNU 2-03a	26	a **6** × 2 = 12 b 5 × **5** = 25 c 7 × 10 = **70** d 2 × **8** = 16 e **10** × 8 = 80 f 5 × **4** = 20 g 35 = **5** × 7 h 100 = 10 × **10**	**On track** • Can accurately complete each calculation. • Is able to use their commutative knowledge of multiplication to help with facts they cannot instantly recall, e.g. initially unsure of 6 × 2 but knows this is same as 2 × 6. **Further learning required** • May not know/recall certain multiplication facts and so calculates inaccurately. • May not understand the commutative property of multiplication. • May struggle with examples where the unknown is a number other than the product (e.g. * × 2 = 12; 5 × * = 20) and wrongly believe the numbers given must always be multiplied together, i.e. attempts to calculate 2 × 12 and 5 × 20. • May not be confident answering questions where the product is at the beginning of the calculation, e.g. questions g and h.

Topic–benchmarks/ Es & Os	Question	Answer	Marking guidance
MNU 2-03a	27	a **2** $2 \times 3 = 6, 3 \times 2 = 6, 6 \div 3 = 2, 6 \div 2 = 3$ b **7** $7 \times 3 = 21, 3 \times 7 = 21,$ $21 \div 3 = 7,$ $21 \div 7 = 3$ c **27** $3 \times 9 = 27, 9 \times 3 = 27,$ $27 \div 3 = 9,$ $27 \div 9 = 3$	**On track** • Understands the link between multiplication and division, and can accurately complete each triangle and write the corresponding calculations. **Further learning required** • May struggle to complete the multiplication triangles either through being unsure of the three times table or not being clear about how the calculations are presented. • May wrongly believe the two numbers given must always be multiplied together to find the missing number. • May not understand the relationships between multiplication and division (commutativity and/or inverse) and so be unable to write two multiplication and two division number sentences to match each triangle.
MNU 2-03a	28	a The array shows **56**. The array can be partitioned in different ways; for example, it could be divided into $(2 \times 8) + (5 \times 8)$ *or* $(3 \times 8) + (4 \times 8)$. b $7 \times 12 =$ **84** cans Pupil may use $(7 \times 10) + (7 \times 2)$. Others ways are possible.	**On track** • Can confidently divide an array into at least two parts to make calculations easier. • Can link the appearance of an array with the multiplication calculation, e.g. can see an array has 8 rows so therefore understands there will be a link to the eight times table and multiplying by 8. **Further learning required** • May struggle to divide an array into parts, e.g. may 'see' 8×7 as $(8 \times 5) + (8 \times 2)$ but be unsure what to do next. • May be unable to record their arrays as calculations. • May be inaccurate in their calculations.

Topic–benchmarks/ Es & Os	Question	Answer	Marking guidance
MNU 2-02a MNU 2-03a	29	a Amman is correct. $15 \times 100 = 1500$. b Finlay started with 8. $8 \times 10 \times 10 = 800$	**On track** • Response to question a shows that the pupil understands what happens to a number when we multiply it by 100. • Can use their knowledge of multiplication as the inverse of division to complete question b, i.e. by working backwards and dividing 800 by 10 then by 10 again to find the unknown. **Further learning required** • May respond or calculate inaccurately. • May justify their response using an incorrect or ineffective strategy, e.g. 'Add two zeros when multiplying by 100'. • May be unable to work backwards to solve b. May answer 80 000 because they 'see' $800 \times 10 \times 10$.
MNU 2-03a	30	Multiplication square shaded or marked to show that the 4 times table is double the 2 times table.	**On track** • Can confidently link the 4 and the 2 times tables to 'doubles' and 'halves' and give examples. **Further learning required** • May not understand the links between the 4 and the 2 times tables.
MNU 2-03a	31	a Isla will finish all her sweets first (**in 4 days**) as she eats six per day. b Amman's sweets will last the longest (8 days).	**On track** • Can interpret the problem as being about 'times as much' or sharing and make links with multiplication and division. • Understands the link between multiplication and division and can use their knowledge of times tables to solve the problem. • May reason that the more sweets that are eaten each day the less time they will last, and vice-versa. **Further learning required** • May reason that Isla will finish first and Amman's sweets will last the longest but be unable to make links with the appropriate operations/calculations. • May be able to select the appropriate operation but need support to recall the multiplication/division facts required to complete the calculations.

Topic–benchmarks/ Es & Os	Question	Answer	Marking guidance
MNU 2-03a	32	Arrays drawn for: a 3×7 and 7×3 b 9×2 and 2×9 c 4×6 and 6×4	**On track** • Can draw arrays for both multiplication number sentences. • Understands that $3 \times 7 = 7 \times 3$ and that the arrays will look identical when they are rotated. **Further learning required** • May be unable to draw arrays for the given multiplication number sentences. • May draw two identical arrays (before rotation) for each pair of multiplication facts.
MNU 2-03a	33	a **8 bags** as $32 \div 4 = 8$ b If there are 8 cupcakes in a bag, the answer will be **halved** as the number of cupcakes in each bag has doubled.	**On track** • Can visualise the problem as making equal groups and make links with multiplication/division. • Understands the relationships between doubling and halving. **Further learning required** • May not recognise the problem as being about 'equal groups' and is therefore unable to visualise, represent or solve it using concrete materials, drawings or known facts. • May not understand the inverse relationship between doubling and halving.
Multiples, factors and primes MTH 2-05a	34	a **true** b **false**: 5 is a factor of 15 *or* 7 is a factor of 14 (other answers are possible). c **true** d **false**: the first three multiples of 3 are 3, 6, 9 *or* the 4th, 5th and 6th multiples of 3 are 12, 15 and 18 (other answers are possible).	**On track** • Understands the vocabulary used in the expressions; multiples and factors. • Can answer accurately and are confident to change the wording of the false expressions so that they are true. **Further learning required** • May be unfamiliar with the words 'multiples' or 'factors'. • May know which expressions are false but may struggle to change them so that they are true.
Fractions, decimal fractions and percentages MNU 2-07a	35	The words that should be written on the fraction wall are: **one half** on each half, **one third** on each third, **one quarter** on each quarter, **one fifth** on each fifth, **one sixth** on each sixth and **one eighth** on each eighth.	**On track** • Can complete the fraction wall with the correct fraction words. Although accurate spelling should be encouraged, reasonable and recognisable attempts (e.g. quater) should be accepted. **Further learning required** • Can complete some but not all of the fraction wall, e.g. more familiar fraction words such as half, quarter and third may be included but not the others.

Topic–benchmarks/ Es & Os	Question	Answer	Marking guidance
MNU 2-07b	36	$\mathbf{\frac{1}{2}} = \frac{2}{4}$ and $\frac{4}{8}$ $\mathbf{\frac{1}{3}} = \frac{2}{6}; \frac{2}{3} = \frac{4}{6}$ $\mathbf{\frac{1}{4}} = \frac{2}{8}; \frac{2}{4} = \frac{4}{8}$ Other answers are possible; for example, some pupils may choose to show equivalence using the whole, e.g. $\frac{5}{5} = \frac{6}{6}$	**On track** • Can write five pairs of equivalent fractions using their completed fraction wall in words or using fraction notation. • Can use the = sign to show equivalence. **Further learning required** • Can name the fractions but has difficulty identifying all but the most common equivalences, e.g. one half = two quarters.
MNU 2-07b	37	Any four from: $\frac{1}{2} = \frac{2}{4}$, $\frac{1}{2} = \frac{3}{6}$, $\frac{2}{4} = \frac{3}{6}$ $\frac{1}{3} = \frac{2}{6}$ $\frac{2}{3} = \frac{4}{6}$ Some pupils may also choose to show equivalence using the whole, e.g. $\frac{6}{6} = \frac{3}{3}$	**On track** • Can manipulate the examples given to create four pairs of equivalent fractions. **Further learning required** • May be unable to create equivalent fractions beyond $\frac{1}{2} = \frac{2}{4}$. • May want to take a piece away from one of the wholes to create 'equivalence', e.g. remove one of the thirds to make 'halves' but with a smaller whole.
MNU 2-07a MNU 2-07c	38	a $\frac{6}{12} = \frac{1}{2}$ b $\frac{5}{15} = \frac{1}{3}$ c $\frac{12}{16} = \frac{3}{4}$ d $\frac{12}{20} = \frac{3}{5}$	**On track** • Can accurately count the number of equal parts coloured and write this as a fraction. • Can write each fraction identified in its simplest form. **Further learning required** • May be able to identify that, for example, six of twelve equal parts/six twelfths are shaded but struggle to write this using fraction notation. • May correctly identify the fractions $\frac{6}{12}$, $\frac{12}{16}$, $\frac{5}{15}$, $\frac{12}{20}$ but be unable to simplify them. • May simplify each fraction (e.g. $\frac{6}{12} = \frac{3}{6}$) but be unfamiliar with the phrase 'simplest form'.

Topic–benchmarks/ Es & Os	Question	Answer	Marking guidance
MNU 2-07b	39	a Accept any one of: $2\frac{1}{2}$; two (whole pizzas) and a half b Accept any one of $3\frac{3}{5}$; three (whole cheesecakes) and three fifths; 18 fifths	**On track** • Can write the fraction represented by each picture as a mixed number or in words. **Further learning required** • May not link the picture with the correct fraction. • May be unable to write the fraction in different ways or does so incorrectly, e.g. may write 1.5 and 1.5 tenths.
MNU 2-07a MNU 2-07c	40	a 1.5, 9.8, 19.5, 19.9, 20.1 b Decimal fractions between the numbers on the list can be: Any fraction larger than 1.5 and smaller than 9.8; Any fraction larger than 9.8 and smaller than 19.5; Any fraction larger than 19.5 and smaller than 19.9; Any fraction larger than 19.9 and smaller than 20.1.	**On track** • Can order the decimal fractions from smallest to largest. • Can identify and write a decimal fraction that lies between each pair of decimal numbers on the number line. • More confident pupils may choose decimal fractions that are slightly further from a pair of decimal fractions on the number line, e.g. they may choose 12.4 as a decimal fraction between 9.8 and 19.5 instead of 9.9 or 19.4. **Further learning required** • May disregard the digit after the decimal point, so believe that 19.9 and 19.5 are equal. • May order the numbers according to the digit in the tenths place only and so believe that 20.1 is the smallest number because it only has one tenth. • May incorrectly identify and write decimal fractions that lie between each pair of decimal fractions on the number line, e.g. they may write 19.10 as the decimal fraction after 19.9.

Topic–benchmarks/ Es & Os	Question	Answer	Marking guidance
MNU 2-07a	41	a 12 km. Bar model should show the whole (27 km) and the parts; $\frac{4}{9}$ (12 km) and $\frac{5}{9}$ (15 km). b 12 girls Bar model should show the whole (32 children) and the parts $\frac{5}{8}$ (20) boys and $\frac{3}{8}$ (12) girls.	**On track** • Can understand and visualise the word problems. • Can represent the information given in the word problem as a bar model. • Can identify the whole and can mentally calculate fractions of amounts to find the parts. **Further learning required** • May struggle to understand and/or visualise the word problem to then represent this as a bar model. • May omit parts of the calculation or add instead of subtracting/counting on, e.g. thinks $\frac{4}{9}$ of 27 km (12 km) is the answer or adds 27 km and $\frac{4}{9}$ of 27 km. • May attempt to calculate fractions of an amount incorrectly, e.g. calculate $\frac{4}{9}$ by dividing by 9 but not multiplying this by 4.
Money MNU 2-09a	42	a £4.27 b £0.75 c £10.02 d £1.00 or £1 e £40.00 or £40 f £0.70	**On track** • Can write each amount correctly using the £ sign and decimal notation. **Further learning required** • May write some or all amounts incorrectly, e.g. 75p might be written as £7.5. • May include both £ and p in responses, e.g. £4.27p. • May be confused as to where to place the decimal point or wrongly believe zero can be omitted, e.g. ten pounds and two pence written as £10.2.

<table>
<tr><th>Topic–benchmarks/ Es & Os</th><th>Question</th><th>Answer</th><th>Marking guidance</th></tr>
<tr><td>MNU 2-09c</td><td>43</td><td>Amman made a loss of £20</td><td>On track
• Can understand and visualise the word problem.
• Can use the information given in the word problem to calculate the answer.
• Can use mental maths strategies or informal jottings to calculate the difference between the original cost of the tickets and how much Amman sold them for.
• Can state whether Amman made a profit or loss.
Further learning required
• May not understand and/or be unable to visualise the word problem.
• May be unable to extract the correct information from the problem.
• May be unable to accurately calculate the answer.
• May use 'past' and 'to' incorrectly.</td></tr>
<tr><td>Time
MNU 2-10a</td><td>44</td><td>a 7:34am
twenty-six minutes to eight (in the morning)
seven thirty-four (accept 0734)
b Hands drawn on clock face to show 3:52
eight minutes to four (in the afternoon)
three fifty-two (accept 1552)
c 12:18am
eighteen minutes past twelve (in the morning/ while I'm sleeping)
twelve eighteen (accept 0018)</td><td>On track
• Can confidently convert between 12 hour and 24 hour time and write the times in words using 'to' and 'past'.
• Can use a.m., p.m. and 24-hour notation where appropriate.
Further learning required
• May confuse 12-hour and 24-hour time notation, especially for times around midday and midnight (question c).
• May use 'past' and 'to' incorrectly.</td></tr>
<tr><td>MNU 2-10a</td><td>45</td><td><table><tr><th>12-hour time</th><th>24-hour time</th></tr><tr><td>midnight</td><td>00:00</td></tr><tr><td>3:20 a.m.</td><td>03:20</td></tr><tr><td>6:30 a.m.</td><td>06:30</td></tr><tr><td>noon</td><td>12:00</td></tr><tr><td>3:27 p.m.</td><td>15:27</td></tr></table></td><td>On track
• Can confidently convert between 12-hour and 24-hour times.
• Can use a.m., p.m. and 24-hour notation correctly.
Further learning required
• May confuse 12-hour and 24-hour time notation, e.g. writes 15:27 p.m.
• May be unfamiliar with the words midnight and noon, refer to 12 a.m. and 12 p.m. and/ or may not know that a new day starts at midnight (00:00).</td></tr>
</table>

Topic–benchmarks/ Es & Os	Question	Answer	Marking guidance
MNU 2-10a	46	a quarter past 11 b quarter to 10 c quarter to 1 d Isla's tv programme finished at 3:30/1530, 3.30pm and half past three/three thirty.	**On track** • Can use the terms 'quarter to' and 'quarter past' accurately. • Can write a.m. and p.m. times in digital and analogue format and in words. • Knows there are 60 minutes in an hour and can solve word problems featuring three quarters of an hour. **Further learning required** • May confuse 'quarter to' and 'quarter past'. • May struggle with duration examples that involve bridging an hour (i.e. 3 o'clock) and so is unable to work out when the programme will finish. • May attempt to solve the problem by incorrectly using a formal written algorithm instead of mental strategies and/or informal jottings.
MNU 2-10a	47	a 25 minutes b 30 minutes c 20 minutes	**On track** • Can accurately read a 24-hour timetable. • Can use 24-hour time to calculate durations that bridge an hour. **Further learning required** • May have difficulty interpreting a 24-hour timetable. • May be unable to bridge the hour when calculating duration, e.g. may attempt to incorrectly use a formal written algorithm instead of mental strategies and/or informal jottings.

Topic–benchmarks/ Es & Os	Question	Answer	Marking guidance
Measurement MNU 2-11a	48	2 cm = 20 mm 4 cm = 40 mm 5 cm = 50 mm 7 cm = 70 mm 10 cm = 100 mm 12 cm = 120 mm	**On track** • Can make a reasonable estimate of the length of each line and can write these estimations in centimetres. They may use their own internal benchmarks to estimate, e.g. I know that my rubber is about 5 cm long. • Can measure the lines accurately with a ruler and write these measurements in centimetres (cm). • Can use their knowledge of 10 mm = 1 cm to convert their measurements into millimetres. **Further learning required** • May make unreasonable estimates. • May have difficulty using a ruler to accurately measure the length of each line; for example, they may not start from the 'zero' point on the ruler or use an incorrect scale to measure in centimetres. • May not know that 1 cm = 10 mm and/or be unable to convert between the two units.
MNU 2-11b	49	Apples = 400 g Carrots = 1.5 kg or 1500 g Backpack = 3 kg or 3000 g	**On track** • Can use their knowledge of kilograms and place value to calculate the graduations on the scales. • Can read the scales accurately and write the mass of each object using the correct unit of measurement, i.e. grams or kilograms. **Further learning required** • May not know how to calculate or understand the value of each graduation on the scales. • May struggle to read the scales or record the mass of each object correctly. • May confuse grams and kilograms.

<table>
<tr><th>Topic–benchmarks/ Es & Os</th><th>Question</th><th>Answer</th><th>Marking guidance</th></tr>
<tr><td>MNU 2-11c</td><td>50</td><td>
<table>
<tr><th>Shape</th><th>Fewest no. of measure-ments</th><th>Peri-meter</th></tr>
<tr><td>rectangle</td><td>2</td><td>10 cm</td></tr>
<tr><td>square</td><td>1</td><td>12 cm</td></tr>
<tr><td>rhombus</td><td>1</td><td>14 cm</td></tr>
<tr><td>hexagon</td><td>1</td><td>15 cm</td></tr>
<tr><td>isosceles triangle</td><td>2</td><td>11 cm</td></tr>
</table>
</td><td>On track
• Understands that perimeter means 'all the way round the edge'.
• Can use their knowledge of the properties of 2D shapes to work out the fewest number of measurements needed to find the perimeter, e.g. knows that a square has four equal sides, therefore, they only need to measure one side and multiply by four.
• Can measure accurately with a ruler, carry out the required multiplications and record the perimeter of each shape in centimetres.
Further learning required
• May be unfamiliar with the word perimeter or confuse perimeter and area.
• May be unsure of the properties of 2D shapes and so rely on measuring each side individually.
• May have difficulty using a ruler to accurately measure the length of the sides. This may be because they do not start from the 'zero' point on the ruler or are using an incorrect scale to measure in centimetres.
• May struggle to multiply some of the measurements or be unsure about effective strategies to calculate the perimeter, especially for calculations involving 0.5 cm.</td></tr>
<tr><td>MNU 2-11c</td><td>51</td><td>Dark grey square (sides of 4 cm) = 16 cm^2
Pale grey rectangle (sides of 2 cm and 4 cm) = 8 cm^2
Debit card (sides of 6 cm and 4 cm) = 24 cm^2
Playing card (sides of 3 cm and 4 cm) = 12 cm^2</td><td>On track
• Can use their knowledge of the properties of 2D shapes to find the surface area of each, e.g. knows that a square has four equal sides, therefore can multiply the length of one side by itself.
• Can measure accurately with a ruler and write these measurements in centimetres (cm).
• Can multiply accurately to calculate surface area and use cm^2 to record this.
Further learning required
• May be unfamiliar with the word area or confuse area and perimeter.
• May be unsure of the properties of 2D shapes and so rely on measuring each side individually.
• May have difficulty in using a ruler to accurately measure the length of the sides. This may be because they do not start from the 'zero' point on the ruler or are using an incorrect scale to measure in centimetres.</td></tr>
</table>

Topic–benchmarks/ Es & Os	Question	Answer	Marking guidance
			• May struggle to multiply some of the measurements or be unsure about effective strategies to calculate the surface area.
MNU 2-11b	52	a 14 ml b 21 ml c 500 ml d 300 ml e 1800 ml, 1 l 800 ml or 1.8 litres	**On track** • Can use their knowledge of millilitres and place value to calculate the graduations on the scales. • Can read the scales accurately and record each measurement using the correct unit: millilitres or litres. **Further learning required** • May not know how to calculate or understand what the graduations on each scale shows. • May only be able to calculate the capacity of liquid in containers where one graduation equals one millilitre. • May confuse millilitres and litres.
MNU 2-11c	53	a 27 cm^3 b 16 cm^3 c 24 cm^3 d 64 cm^3 e 10 cm^3	**On track** • Can accurately recreate the models shown with centimetre cubes. • Can use the dimensions, or count the cubes, to calculate the volume. • Can record the volume of each cuboid using the words 'cubes', 'cubic centimetres' or the notation cm^3. **Further learning required** • May be unable to recreate the models using centimetre cubes. • May struggle to count the cubes to calculate the volume, e.g. may need to break the model down into smaller parts to find the volume of the whole cuboid. • May not connect 'cubes', 'cubic centimetres' or cm^3 with volume.
Patterns and relationships MTH 2-13a	54	a **moon, three stars** – three stars and a moon is the pattern b **R P** – starting at Z, write the alphabet missing every other letter c **14, 5** – subtract 9 d **22, 29** – add 1, add 2, add 3…add 6 and add 7 for the two answers.	**On track** • Can identify, continue and explain each pattern. **Further learning required** • May think that patterns cannot contain anything other than numbers or shapes, or that a pattern cannot have two of the same thing next to each other. • May have difficulty continuing the pattern. • May be unable to describe the pattern in writing.

Topic–benchmarks/ Es & Os	Question	Answer	Marking guidance
Expressions and equations MTH 2-15a	55	a 8 × **?** = 32 b **?** ÷ 4 = 6 c 28 ÷ 4 = **? or** 4 × **?** =28	**On track** • Can visualise and understand each word problem. • Can write an equation to match each word problem. **Further learning required** • May have difficulty interpreting the word problems. • May be unable to represent each word problem as an equation, e.g. may not be able to insert the unknown ('?') in the correct place.
2D shapes and 3D objects MTH 2-16a MTH 2-17b	56	a A square is drawn with sides of 5 cm b A rectangle is drawn with sides of 3 cm and 4 cm	**On track** • Can competently use the protractor to measure 90 degrees. • Can use a ruler to draw straight sides of a given length. • Can coordinate both of these steps to produce a square and rectangle of a given size. **Further learning required** • May not realise that the sides of each shape must be perpendicular. • May know they must create four right 90-degree angles but cannot work out how to use a protractor to do this. • May have difficulty using a ruler in coordination with the protractor to draw the 2D shapes, e.g. may just use the ruler and estimate the 90-degree angle or use the corner of the ruler to produce this.
MTH 2-16a	57	Has straight sides – **rectangle, rhombus, hexagon, trapezium, kite** sketched. Has exactly four sides – **rectangle, rhombus, trapezium, kite** sketched. Has pairs of parallel sides – **rectangle, rhombus, hexagon** sketched. Has square vertices – **rectangle** sketched. Other – **circle** sketched.	**On track** • Can understand the vocabulary used to sort the shapes. • Can sort the shapes accurately according to these properties. **Further learning required** • May struggle to recognise shapes that are not orientated along a horizontal line, e.g. the rectangle. • May be confused about the vocabulary that is used.

Topic–benchmarks/ Es & Os	Question	Answer	Marking guidance
MTH 2-16c	58	Top, one side and one end view of each shape drawn on isometric paper (answers may vary according to end and side viewed).	**On track** • Can understand the vocabulary related to drawing 3D objects. • Can visualise the 3D object represented by the 2D drawing. • Can identify top, end and side views of the 3D object. **Further learning required** • May be unable to draw the top, end or side view of the 3D object, as they cannot visualise the 3D object and/or do not understand the vocabulary.
MTH 2-16a	59	At least one triangular face – empty. Intersection – **B** (square-based pyramid), **C** (triangular prism), **D** (pentagonal based pyramid). Five or more vertices – **A** (cube), **E** (hexagonal prism).	**On track** • Can use the Venn diagram to sort their answers so that it is accurate. **Further learning required** • May not know what a Venn diagram is. • May be unable to accurately sort the 3D objects into the appropriate sections of the Venn diagram, e.g. may not use the intersection to show 3D objects that have both properties.
Angles, symmetry and transformation MTH 2-17a	60	Acute angles (3) – C, D, H Right angles (3) – A, F, J Obtuse angles(4) – B, E, G, I	**On track** • Can recall the vocabulary of angles and properties of acute, right and obtuse angles. • Can identify acute, right and obtuse angles on the house diagram and record these angles accurately in the table. **Further learning required** • May confuse acute, right and obtuse angles and their properties. • May be unable to identify acute, right and/ or obtuse angles by sight or lack strategies to help them, e.g. rotating the paper to look at the angle from a different perspective or using a 'right angle tester'.
MTH 2-17c	61	A variety of answers will be given depending on the size of the 'F' they have drawn and their starting point.	**On track** • Can sketch the letter 'F' and provide directions using compass points, e.g. W2, S4. **Further learning required** • May confuse compass points, most commonly east and west.

Topic–benchmarks/ Es & Os	Question	Answer	Marking guidance
MTH 2-18a	62	a The square has vertices at the following co-ordinates: (1,1), (3,1), (1,3), (3,3) b A rectangle should be drawn with vertices (6,6), (1,6), (6,4) and (1,4).	**On track** • Understands that vertices means corners. • Can accurately record compass points plotted on a grid. • Can plot the points given and join them to create a rectangle. **Further learning required** • May not understand the word vertices. • May confuse the horizontal and vertical axes, e.g. writes (6,1) instead of (1,6).
MTH 2-19a	63	Horizontal line of symmetry – B, E, G. Vertical line of symmetry – A, C, H. No line of symmetry – D, F.	**On track** • Can identify lines of symmetry in the 2D shapes given and draw these using a ruler to make a broken line. • Understands the difference between horizontal and vertical symmetry. **Further learning required** • May confuse the words horizontal and vertical. • May not draw the lines of symmetry accurately.
MTH 2-19a	64	Answers will vary depending on the design chosen. Flag should have one vertical line of symmetry and include four different colours.	**On track** • Can create a flag with four colours and one line of vertical symmetry. **Further learning required** • May be correct in one part of the flag (e.g. nearest the line of symmetry) but be unable to reflect the entire pattern.
MTH 2-17b	65	a 20° b 90° c 45° d 160° and e 135°	**On track** • Can use a protractor to measure acute, right and obtuse angles accurately within five degrees. • Knows to position the protractor with the zero line along one arm of the angle, the vertex where the arms of the angle join and reads the correct scale. **Further learning required** • May not start at the 'zero line', positions the vertex incorrectly or uses the wrong arm to take the measurement, e.g. reads an acute angle measuring 45 degrees as a 135 degree angle.

Topic–benchmarks/ Es & Os	Question	Answer	Marking guidance
MTH 2-17d	66	a 100 km (2.5 cm) b Oban c Ullapool and Dumfries (difference of 240 km/6 cm on the map) d Perth and Dundee (difference of 20 km/0.5 cm on the map)	**On track** • Can confidently use a ruler and a simple scale to calculate distance. • Can use effective mental strategies or informal jottings to calculate distances. **Further learning required** • May use a ruler incorrectly, e.g. starting at one instead of zero. • May not understand how to use calculate actual distance from a scale • May be unable to choose an appropriate method to multiply by 40.
Data handling and analysis MNU 2-20a	67	a 160 b 20 c Day 4 d Days 4 & 5	**On track** • Can confidently read and interpret the pictograph. • Can use effective mental strategies and informal jottings to make accurate calculations. **Further learning required** • May be unable to interpret the pictograph, e.g. may fail to observe the key and believe one image represents one mobile phone. • May not understand the language used in the questions or interpret the questions incorrectly.
MNU 2-20a	68	a February and June b 300 c April–June 2450 laptops sold, Jan–March 950 laptops sold; therefore, a difference of **1500**.	**On track** • Can confidently read and interpret the line graph. • Can use effective mental strategies and informal jottings to make accurate calculations. **Further learning required** • May struggle to interpret the line graph, e.g. has difficulty understanding the axes and scale on the line graph. • May misinterpret the questions or be unable to manipulate the numbers involved.
MNU 2-20a MNU 2-20b	69	Crisps – **36** Fruit – **30** Sweets – **14** Biscuits – **12** Veg Sticks – **18** 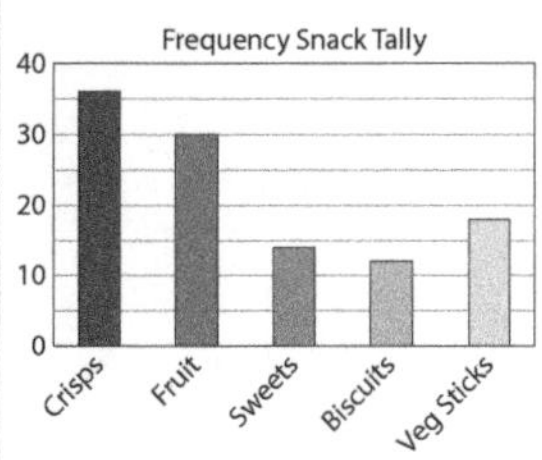	**On track** • Can confidently read and interpret 'tally' and 'frequency'. • Can count tally marks to complete a frequency table and draw a bar graph. **Further learning required** • May struggle to complete an accurate bar graph from the tally marks given. This may be because of inaccurate axes or bars drawn.

Topic–benchmarks/ Es & Os	Question	Answer	Marking guidance
MNU 2-21a	70	a **gymnastics – 12** (as this is $\frac{1}{4}$ of 48) **rugby – 6** (rugby is $\frac{1}{8}$ of the number of children. This is half of a quarter. $\frac{1}{4}$ is 12, therefore $\frac{1}{8}$ will be 6). b rugby = 6, gymnastics = 12. **The difference between them is 6**.	**On track** • Knows there are 360 degrees in a 'full turn' and can use their knowledge of fractions and division to calculate amounts from the pie chart. **Further learning required** • May recognise that $\frac{1}{4}$ of the class like gymnastics and 'half of $\frac{1}{4}$' ($\frac{1}{8}$) like rugby but does not make the link with division. • May struggle to work out the number of children represented in each section of the pie chart, either because of a lack of confidence in fractional understanding or in their methods of dividing. • May be unable to choose a method for finding a fraction of a quantity. • May have difficulty working out smaller sections of the pie chart, e.g. $\frac{1}{8}$ of 48.
Ideas of chance and uncertainty MNU 2-22a	71	Answers will be dependeant on individual responses. Personal statements for: impossible, unlikely, even chance, likely and certain should be given.	**On track** • Can accurately describe the likelihood of simple chance and probability events occurring in their own lives. **Further learning required** • May be unable to or struggle to describe events in their life using the vocabulary of chance and probability and/or give unrealistic examples. • May be able to describe events that are impossible or certain but may have difficulty choosing possible events that have an even chance.

Yearly progress check 2B

Resources

Resources provided online:

Digit cards

Resources needed and available in the setting:

Rulers • Protractors • Isometric dotted paper • Squared paper • Compasses for drawing circles

Questions

1. Place 3756 in its approximate place on the number line below.

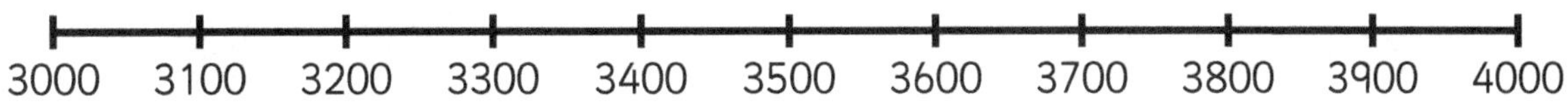

Round 3756 to the nearest 100.

What is 3756 rounded to the nearest thousand?

2. Complete this table to show each decimal number rounded to the nearest tenth and to the nearest whole number.

	Nearest tenth	Nearest whole number
4·56		
101·01		
150·52		

3. Finlay and Nuria are using rounding to help them estimate the answer to 4829 + 3755. Finlay finds the approximate answer by rounding to the nearest 100 and Nuria finds the answer by rounding to the nearest 1000. Whose estimate is closest to the actual answer? Explain your thinking.

Yearly progress check 2B

4. Amman thinks that 80080 says eight hundred and eighty. Explain why he is incorrect.

____

____

Write the number 80080 correctly in words.

____

5. Isla made a number using these place value arrow cards.

What number did Isla make?

____

Amman made a number using these place value counters.

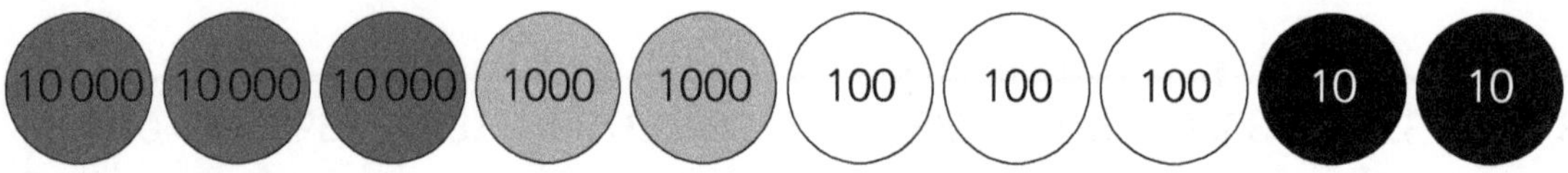

What number has he made?

____

6. Isla is making four-digit numbers using place value counters, but the thousands counters have gone missing! How can she make the following numbers using only 100s, 10s and 1s counters? 

a 2130

b 1004

7. Write the missing numbers.

a 76539, 77539, ______, ______, 80539, 81359, 82359, ______, ______

b 10560, 10550, 10540, ______, ______, 10510, ______, ______, 10480

c 20350, 20250, 20150, ______, ______, ______, 19750, 19650

8. Complete these statements by writing the symbol <, > or = in each box.

a 41649 ☐ 43649

b 27270 ☐ 27027

c 48020 ☐ 48200

9. Write each label's number in numerals.

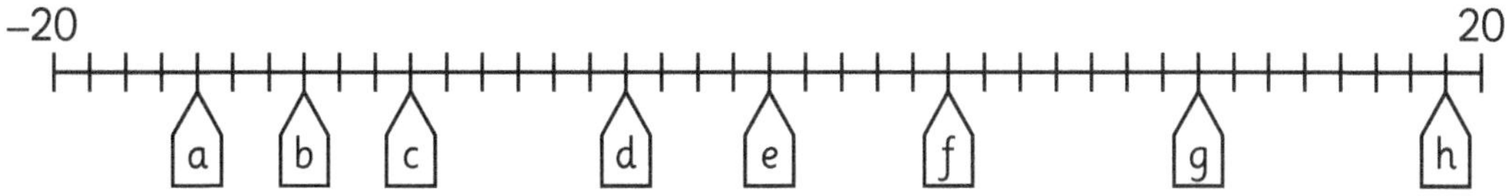

___ ___ ___ ___ ___ ___ ___ ___

Write these temperatures in order from coldest to hottest.

3°C, –49°C, 0°C, 15°C, –12°C

____

Write the coldest and hottest temperature in words using *degrees Celsius.*

____

____

10. Colour each decimal fraction, then write it in words:

a 0·47

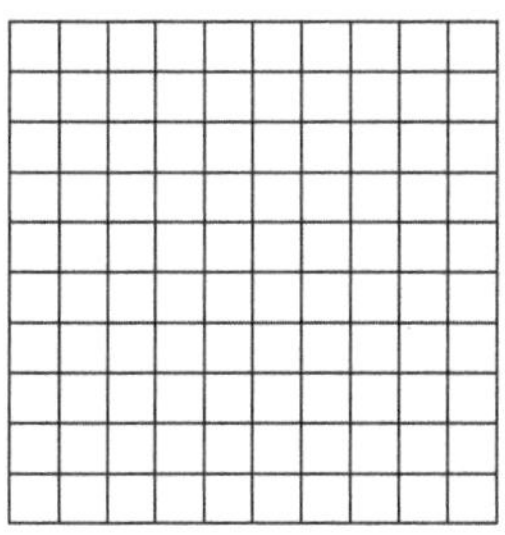

b 1·25

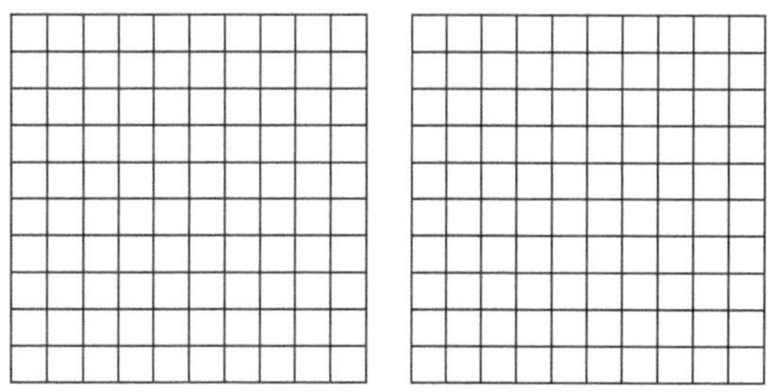

______________ ______________

c 2·03

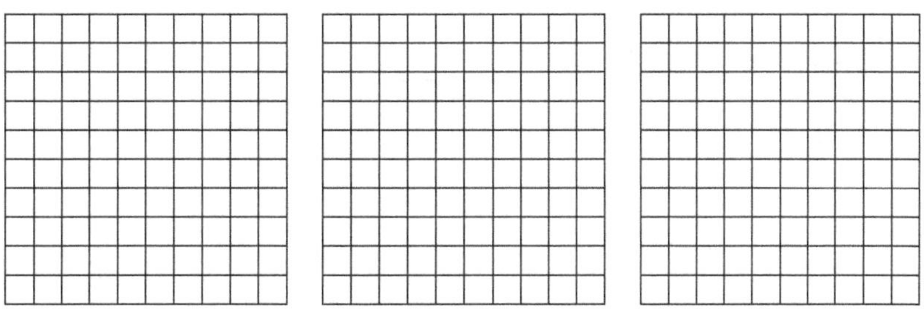

____

Write each decimal fraction as a mixed number.

d 5·46 ________ e 3·03 ________ f 12·98 ________

11. Use the arrow cards to make three different decimal fractions with hundredths.

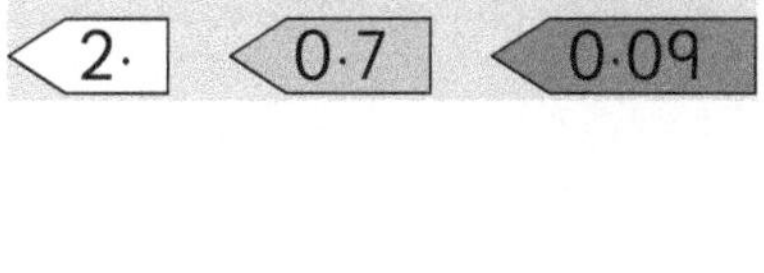

__________ __________ __________

12. Write the decimal fraction that each arrow is pointing to.

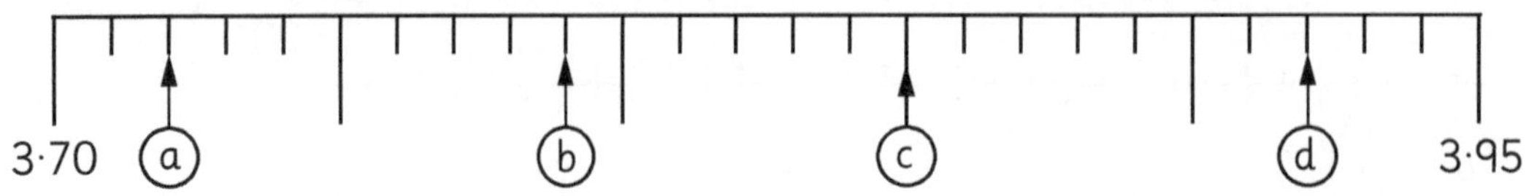

_______ _______ _______ _______

e Write these numbers in order from smallest to largest.

5, 2·8, 28, 5·28, 2·85, 5·82 ____________________

13. Are these statements true or false? Explain your thinking. You may want to use jottings to do this.

a 2852 + 1374 = 2853 + 1375 ____________

b 7999 – 3999 = 8000 – 4000 ____________

c 2789 + 7552 = 7552 + 2789 ____________

d 4503 – 2012 = 4512 – 2003 ____________

14. Calculate mentally using multiples of 10, 100 or 1000. Make jottings to help you keep track of your thinking.

a 229 + 111 + 1238 ____________

b 3000 + 2590 + 410 ____________

c 8000 – 4760 – 2240 ____________

15. Write a multiple of 100 or 1000 in the box to make each number sentence true. Explain how you worked out which number to place in the box.

a 89 119 = ☐ + 86 119

b 54 700 = ☐ – 21 000

16. Use partitioning to calculate the answers to these number sentences.

a 5728 + 4258 = ____________

b 91 766 – 81 510 = ____________

17. Fill in the missing digits.

```
   5 7 8 6          4 7 5 5
 + 2 * 3 *        + * * * *
 ---------        ---------
   * 7 * 7          7 4 4 4
```

18. Finlay has tried to work out the answer to these calculations using column subtraction but has made mistakes. Can you see where he has gone wrong? Write what Finlay should have written.

a

```
  725
 -354
 ----
  400
   30
    1
 ----
  431
```

b

```
  871
 -467
 ----
  400
  -10
   -6
 ----
  384
```

19. Solve this problem using a method of your choice. Show how you worked it out. Nuria's family are planning a holiday costing £867. They have saved £479 already. How much money do they still have to save?

20. Represent this word problem as a bar model, on an empty number line and as a calculation. You may draw a Thinkboard if this helps.

In 1783, Caroline Herschel was the first woman to discover a comet. How many years ago was this?

21. Solve this word problem using strategies of your choice. Explain how you worked it out.

There are 16 000 tickets for the football Cup Final. 7100 tickets are sold in the first week and 3450 are sold in **each of** the next two weeks. How many tickets are left unsold?

22. Calculate mentally. Explain how you worked each answer out.

a 169 + 13·7 ________________

b 51·8 + 20·7 ________________

c 70·7 + 2·4 ________________

d 31·4 + 25·9 ________________

23. The value in the top box is the sum of the two lower boxes. Find the missing numbers. Write two addition facts and two subtraction facts for each bar model.

a

?	
24.9	38.1

____

____

b

81	
60.5	?

____

____

c

100	
?	54.1

____

____

d

?	
79.4	40.6

____

____

24. Show how you could use your knowledge of the 3 times table to work out these calculations involving multiplying by 6. You might want to draw an array to help you.

a 6×2 ______

b 6×5 ______

c 7 groups of 6 ______

d 18 divided by 6 ______

25. What strategies do you know that can help you with the 9 times table? Explain these below. You can use jottings or drawings to help.

26. Isla thinks that 8500 divided by 10 is 85. Amman thinks it is 850. Who is correct? Explain why.

27. Use known facts to help you work out these calculations.

a 360 ÷ 6

Known fact: ☐

So, 360 ÷ 6 = __________

b 2400 ÷ 4

Known fact: ☐

So, 2400 ÷ 4 = __________

c 210 ÷ 30

Known fact: ☐

So, 210 ÷ 30 = __________

28. There are 24 children at an after-school club. They are split into equal teams.

How many teams could be made with:

a six children in each __________

b three children in each __________

c four children in each __________

29. Use brackets to help partition and solve these calculations.

a 378 × 5 __________

b 96 ÷ 4 __________

30. Use a grid to help partition and solve these calculations.

a 432 × 8 __________

b 78 ÷ 3 __________

31. Explain how you can use multiplying by a multiple of 10 to work out 6 × 29.

32. Nuria and Amman were picking strawberries. Nuria picked eight crates of strawberries. Amman picked six crates of strawberries. Each crate contained six punnets.

  a How many punnets did Amman pick? ___

  b How many more punnets did Nuria pick than Amman? ___

33. Work out the answers to the following calculations. You could use a place value grid to help you.

  a 4·7 × 10 = ☐

  b 15·2 × ☐ = 152

  c 3·4 × ☐ = 3400

  d ☐ ÷ 10 = 460

  e 427 ÷ 10 = ☐

  f 9100 ÷ 1000 = ☐

34. a Four friends go to dinner. The bill comes to £38. How much should they each pay if they split the bill equally? Show your working.

  b Nuria runs 3 kilometres over six days. How many kilometres did she run on average each day? Show your working.

Yearly progress check 2B

35. List all the factors for these numbers.

a 8 ____________ b 12 ____________ c 18 ____________

36. Which of these numbers are not multiples of 3? How do you know?

6, 8, 23, 27, 12, 17, 42, 60

____

37. Use the fraction wall to show which fractions can be changed into tenths. Choose from halves, thirds, quarters, fifths, sixths, eighths and twelfths.

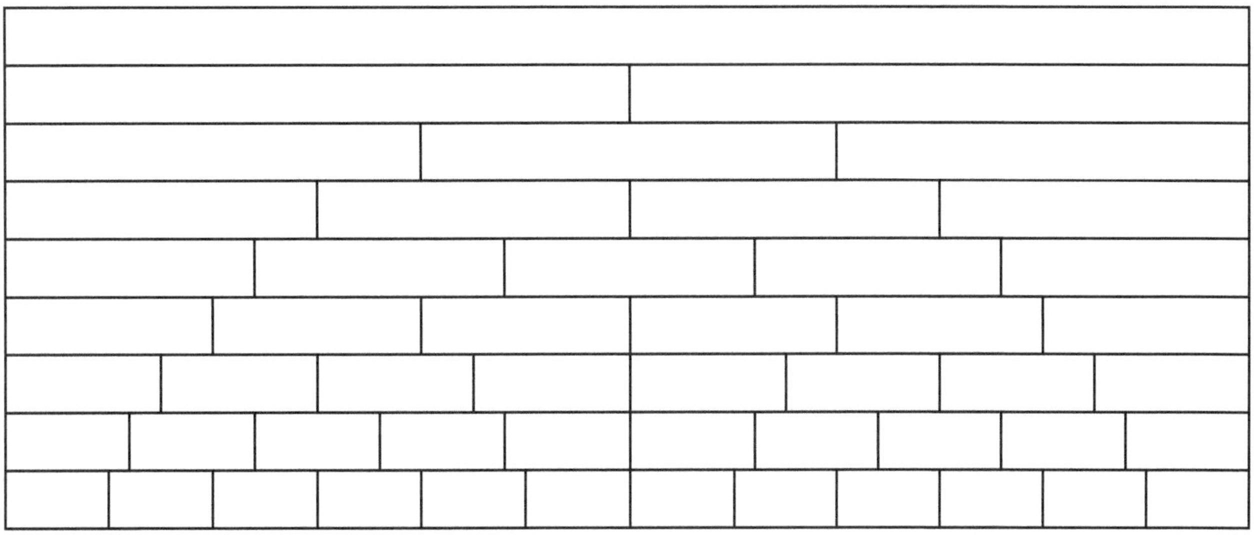

38. Colour the fraction wall to show equivalent fractions to:

a $\frac{2}{3}$ b $\frac{3}{5}$ c $\frac{1}{2}$

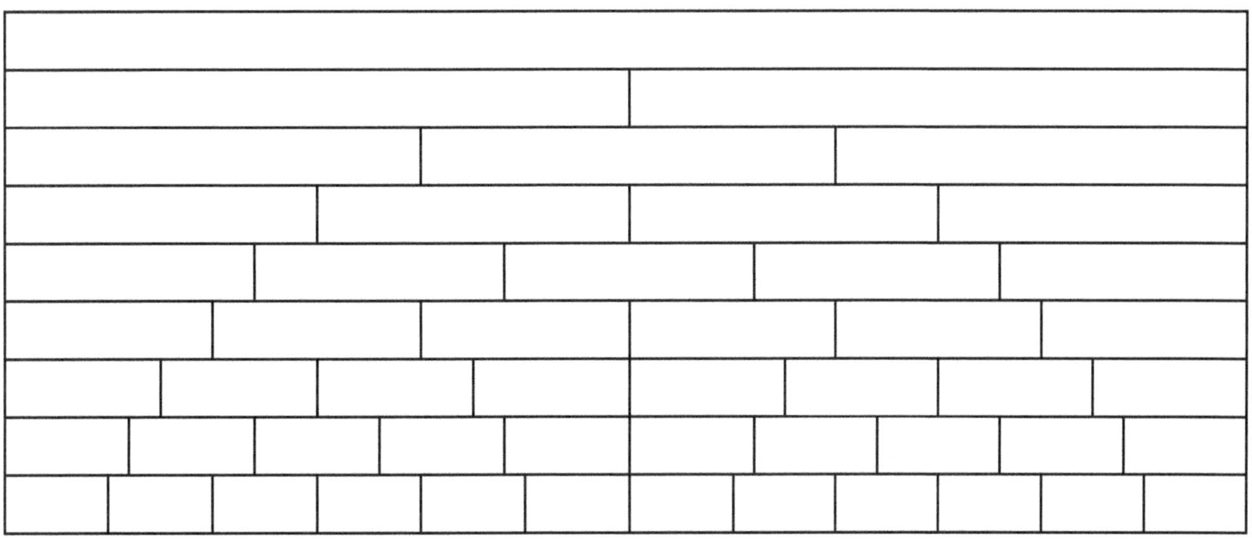

39. a Use equivalence to order these fractions from smallest to largest. Write each fraction in the correct position on the number line.

$\frac{2}{5}$ $\frac{1}{4}$ $\frac{6}{10}$ $\frac{14}{20}$ $\frac{1}{2}$

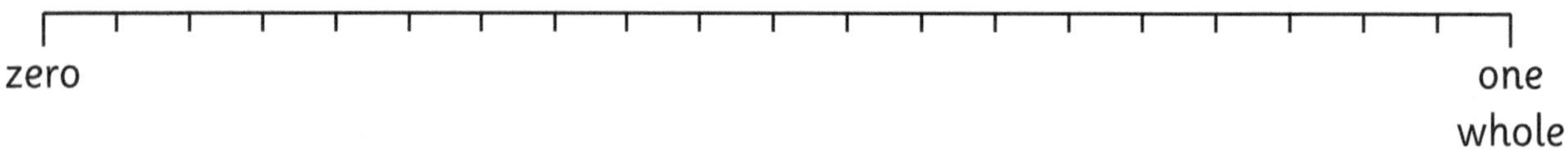

b Isla ate one third of a cake. She cut what was left into six slices and ate two.

Finlay had the same size cake. He ate one half of the cake. He cut what was left into four slices and ate one.

Who ate more cake? Show how you worked this out.

40. Complete the table to show three different ways of writing the same amount.

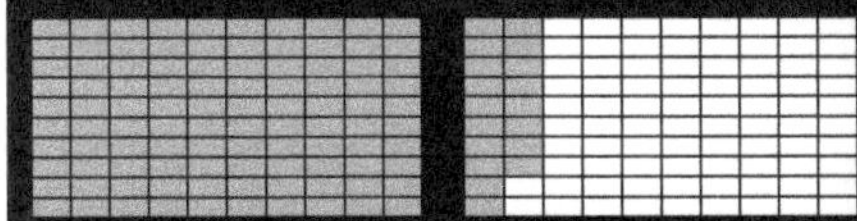	**Whole and hundredths e.g. 1 whole and 18 hundredths**	**Hundredths e.g. 118 hundredths**	**Decimal fraction e.g. 1·18**
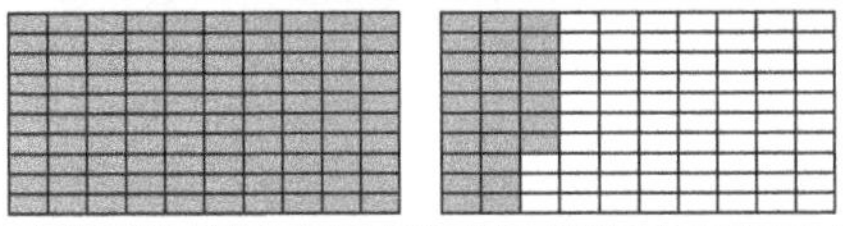			
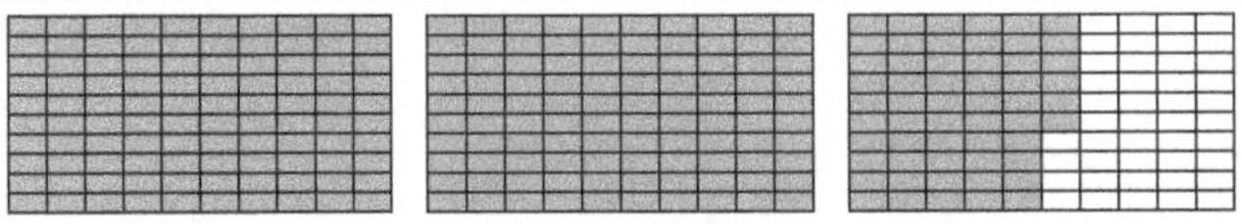			
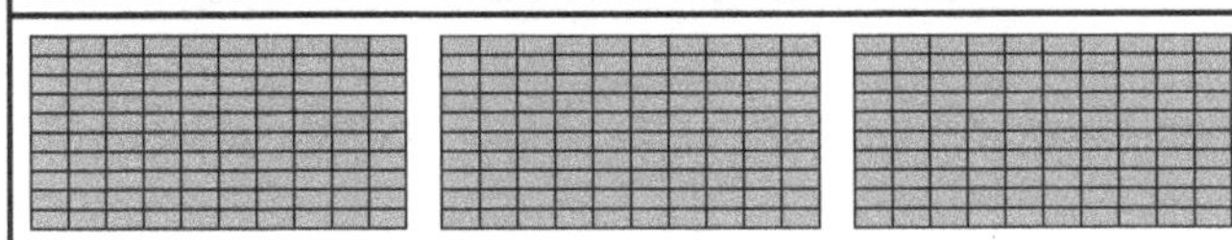			

41. Tick the fractions that can be converted exactly into hundredths and write the equivalent fraction (tenths or hundredths) and the decimal fraction for the fractions you have ticked.

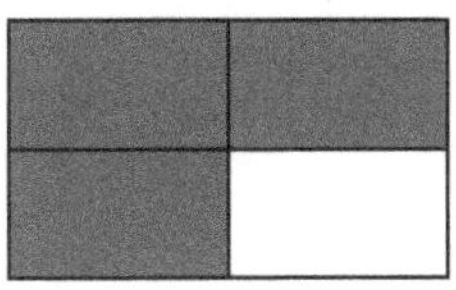

three quarters = ________ = ___

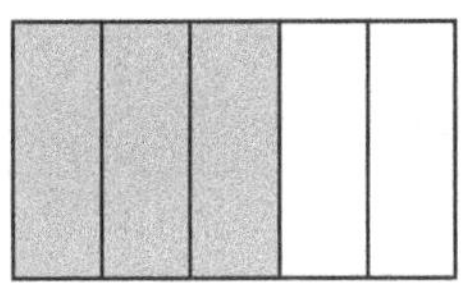

three fifths = ______ = ___

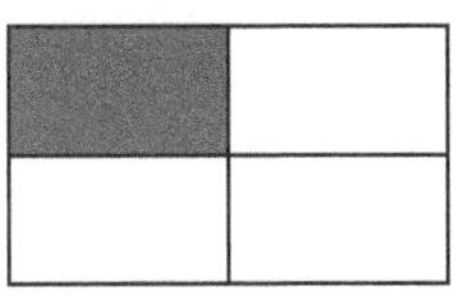

one quarter = ________ = ___

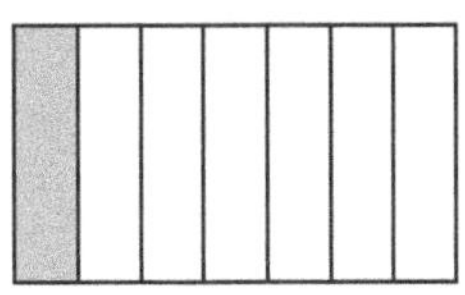

one seventh = ______ = ___

42. Use the bar models to work out the following:

a $\frac{3}{4}$ of 76

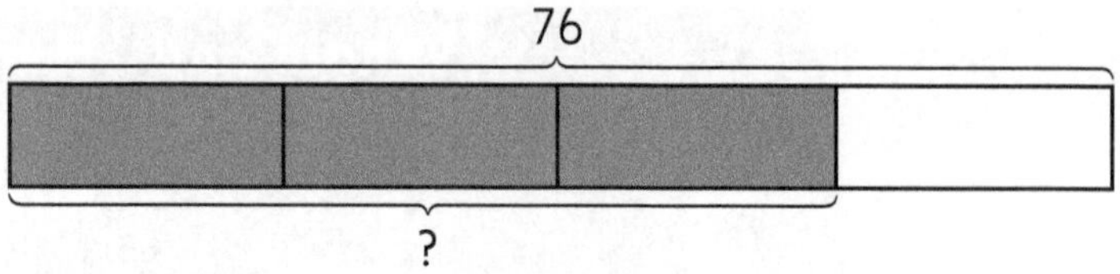

____

b $\frac{5}{6}$ of 132

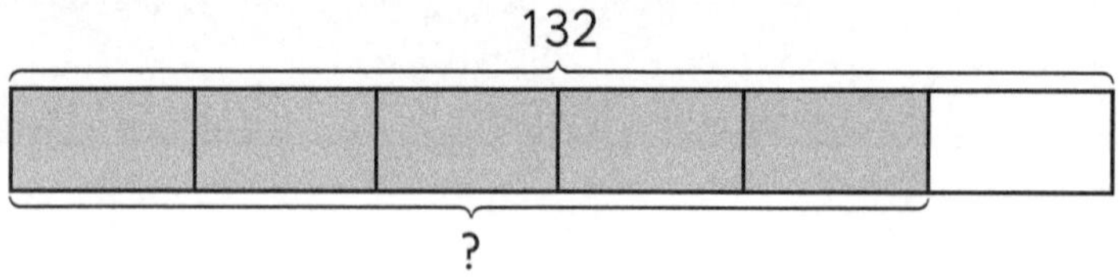

____

c $\frac{7}{10}$ of 5650

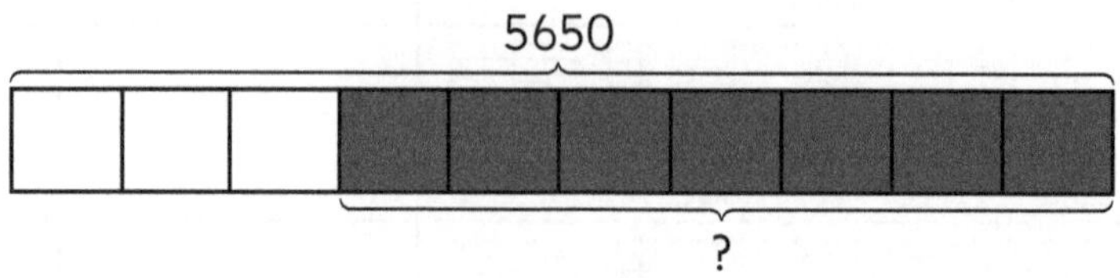

____

d $\frac{5}{8}$ of the population of a village are under the age of 50. A total of 560 people in the village are under

50 years of age. What is the total population of the village? Use the bar model to help you.

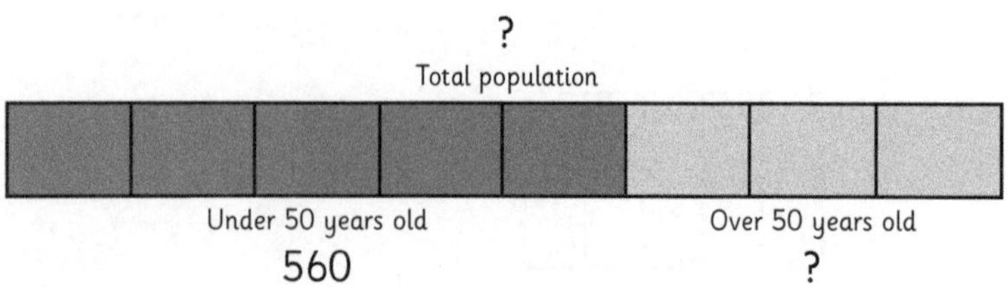

____

Yearly progress check 2B

43. Amman, Finlay, Isla and Nuria were running 200 m to see who was fastest. Their times are recorded in the table below.

Name	Time	Position
Amman	37·16 seconds	
Finlay	38·20 seconds	
Isla	35·05 seconds	
Nuria	35·50 seconds	

Who was first, second, third and fourth?

____

Write a time that would mean that you were placed second if you had joined them in the race.

____

44. Three concerts are taking place. The seating plan shows seats that have been bought (shaded) and seats that are still free (white).

Write the amount of seats bought as a percentage and a fraction, e.g. 1%, 1 out of 100, $\frac{1}{100}$.

a

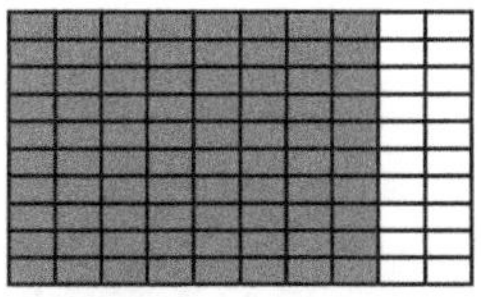

____

b

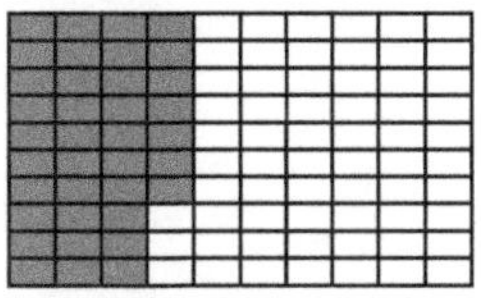

____

c

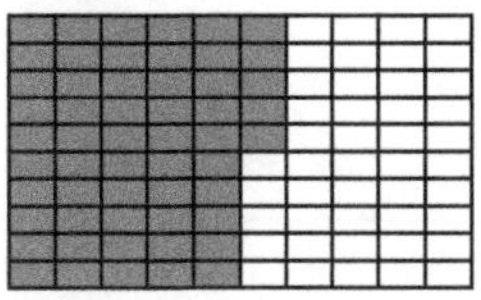

____

d

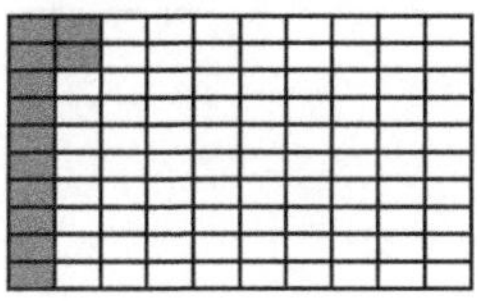

____

Yearly progress check 2B

45. Convert the following fractions to percentages.

a

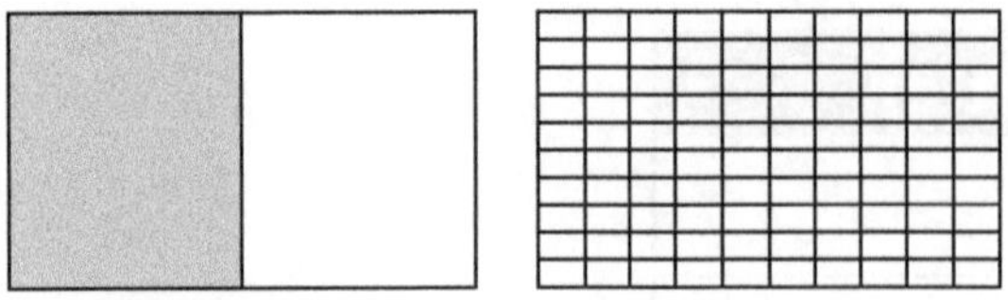

$\frac{1}{2}$ = ______%

b

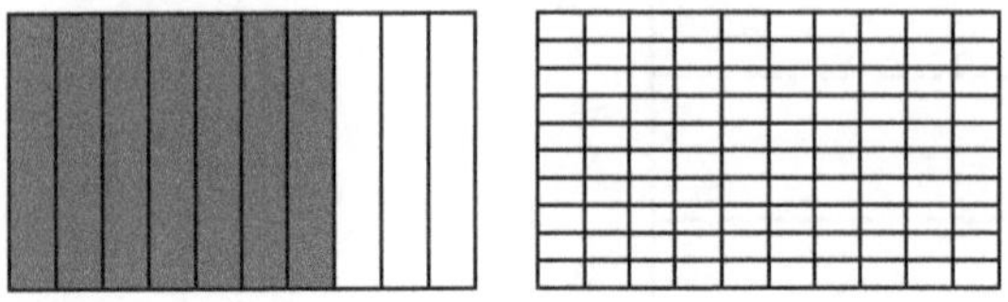

$\frac{7}{10}$ = ______%

c

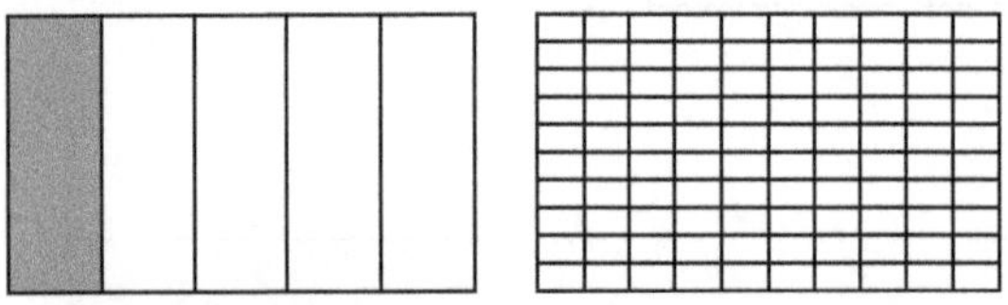

$\frac{1}{5}$ = ______%

d

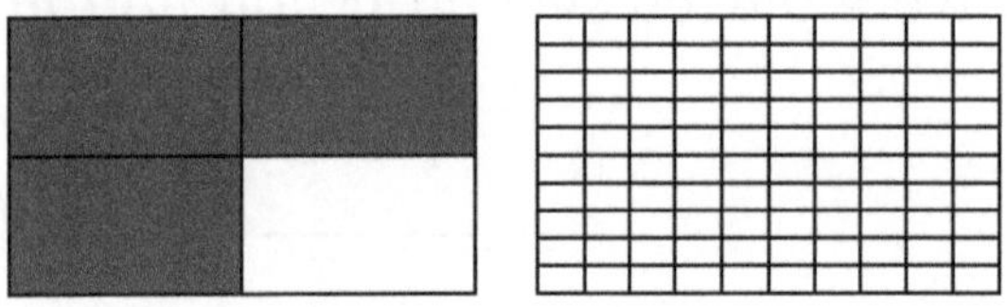

$\frac{3}{4}$ = ______%

46. a A farmer has 120 animals on her farm. 60% of them are cows. How many cows does she have on the farm? You can draw a bar model to help you.

b Write a word problem to match the bar model below.

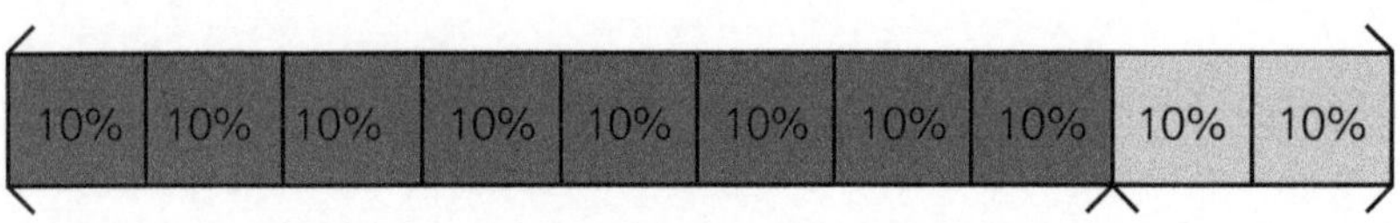

____

____

____

47. Amman and two friends went to the cinema. Amman had some birthday money left, so he paid for all their tickets. Each ticket cost £5·75. Amman had two £5 notes, three £1 coins and nine 50p coins.

a How much money does Amman have? ____________________

b Does he have enough to pay for the three cinema tickets? Explain your thinking.

c Popcorn is 75p. Does Amman have enough to buy some? If not, how much more would he need?

48. Isla has joined a swimming class.

- Individual lessons are £4·50 and the lockers are £1.
- A block of 10 lessons costs £40 but locker hire is free.
- If you pay monthly it costs £16 for four lessons but you need to pay for the lockers.

What is the most Isla might pay for each lesson including locker hire? Show how you worked this out.

49. Nuria's class are running a tuck shop at breaktimes. In the winter, they make a large pot of porridge to sell in small tubs.

a The oats, milk, tubs, spoons and raisins cost them £4 altogether. They can make 20 tubs of porridge from this. What is the least amount of money they should charge so they do not make a loss? Explain your thinking.

b If the class wanted to charge 25p per tub, how much profit would they make if they sold all 20 tubs?

c The class charge 25p per tub but now get a 10% discount on the ingredients. What profit will they now make if they sell all 20 tubs?

50. a Explain the difference between a credit card and a debit card.

____

____

b Write three advantages and three disadvantages of using a credit card.

____

____

____

____

____

____

51. Complete the following table.

12-hour clock	24-hour clock
10:00 a.m.	
	1200
4:30 a.m.	
	1745
12:00 midnight	

52. Finlay catches a bus at 8:25 a.m. On the bus timetable, buses leave every ½ hour after this. What times do the next four buses leave? Write each time in words and in 24-hour time e.g. 0825 and twenty five minutes past eight.

____

____

____

____

53. Using seconds, minutes, hours or days, what is the best unit to measure the time to:

a Read a book ____________________

b Fly to the USA ____________________

c Sneeze ____________________

d Brush your teeth ____________________

54. Finlay, Nuria, Isla and Amman take their bikes to the local track. They decide to have a relay race. The table shows how long each of them took to cycle round the track.

Finlay	2 min 35 s
Nuria	2 min 40 s
Isla	2 min 25 s
Amman	2 min 40 s

How long did their race last in total? Show how you worked it out.

Yearly progress check 2B

55. Estimate the length of each of the bars below in centimetres and millimetres. Write down your estimates in the table. Next use a ruler to measure the actual lengths and record your answers in the table. How close were your estimates? Record the difference between each estimate and measurement in the last column of the table.

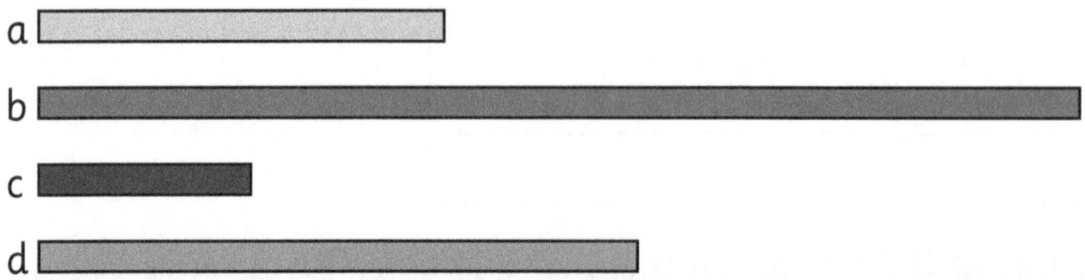

	Actual length (in cm and mm)	Estimate (in cm and mm)	Difference
a			
b			
c			
d			

56. Write the mass of each object in both grams and kilograms in the table below.

a

b

c

What three items in your environment might weigh the same as the three bags above? Write your estimates in the table below.

	Mass in grams	Mass in kilograms and grams	Possible item
a			
b			
c			

57. How much liquid is contained within the measuring jugs below? Write the measurements in litres and millilitres.

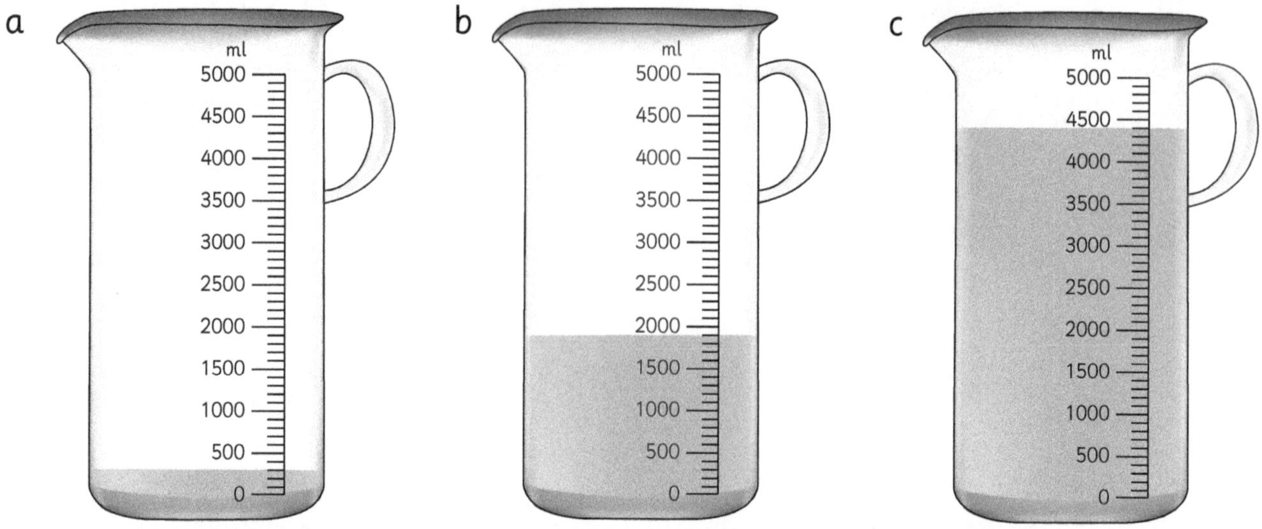

	Capacity in l and ml
a	
b	
c	

58. Sometimes we use imperial measurements. What might be measured using the imperial measurements below?

a stones ______

b feet ______

c miles ______

d pints ______

59. a Draw a square and a rectangle each with a perimeter of 20 cm.

b Calculate the area of the square and rectangle you have drawn. What do you notice about the area of each shape compared to its perimeter?

c Draw a triangle that has half the area of your square and another that has half the area of your rectangle.

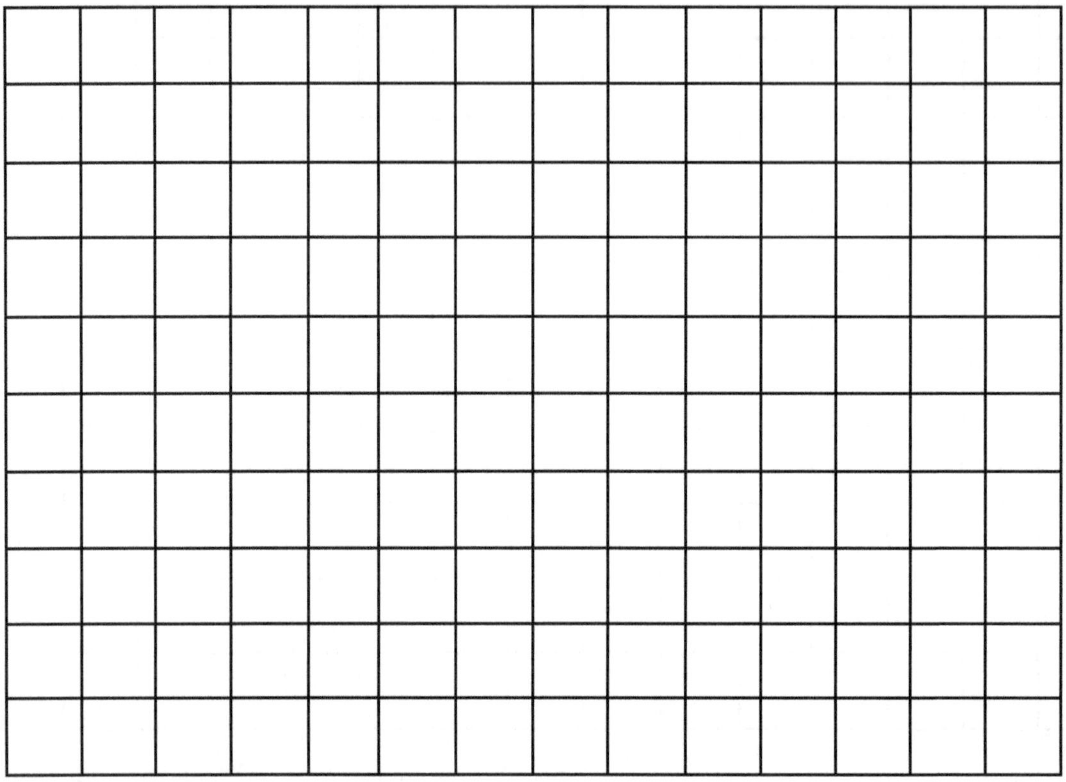

60. Nuria and Isla are building cuboids with centimetre cubes. They have built their first layer.

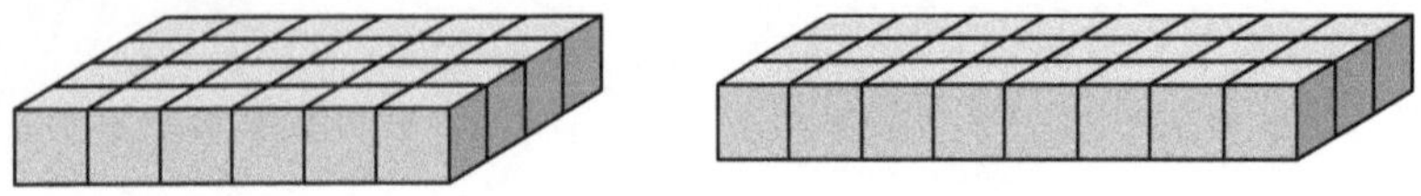

They each have enough cubes to build 5 layers.

a What will the volume of each of their cuboids be?

____

b What do you notice about your answers? Why do you think this is?

61. Complete the table below to help you identify the number pattern..

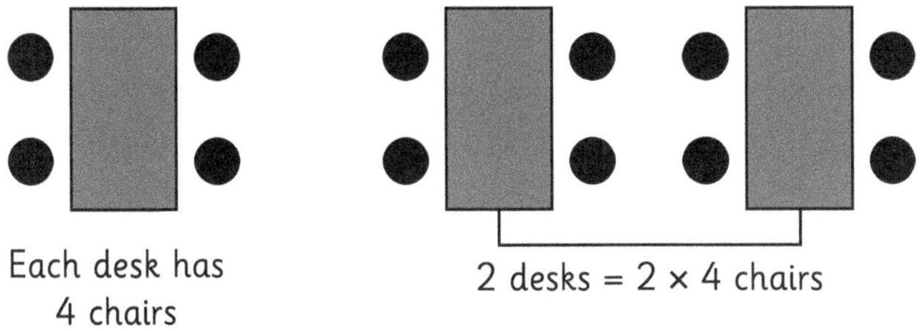

Number of desks (D)	1	2	3	4	5
Number of chairs (C)	4	8			

Complete the formula: Number of chairs = ☐ × number of desks

Now write the formula using the symbols D and C ____________________

How many chairs would you need for 8 desks? How do you know?

62. For each number sentence, find the unknown number.

a $4 \times \square = 24$

b $\square \times 5 = 30$

c $\square \times 4 = 19 - 7$

63. Complete the table then draw the two circles.

Circle	Radius	Diameter
A	5 cm	
B		60 mm

64. Predict what 3D object each net will make when it is folded. Write the name of the shape beside its net.

a 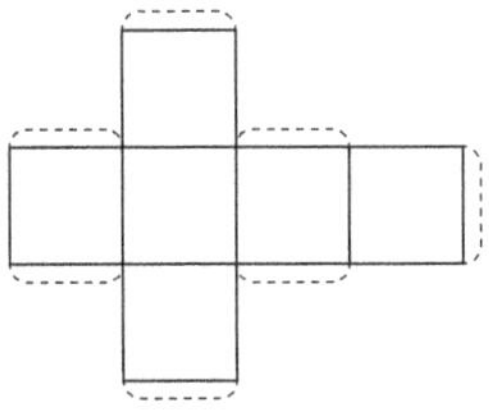

____

b 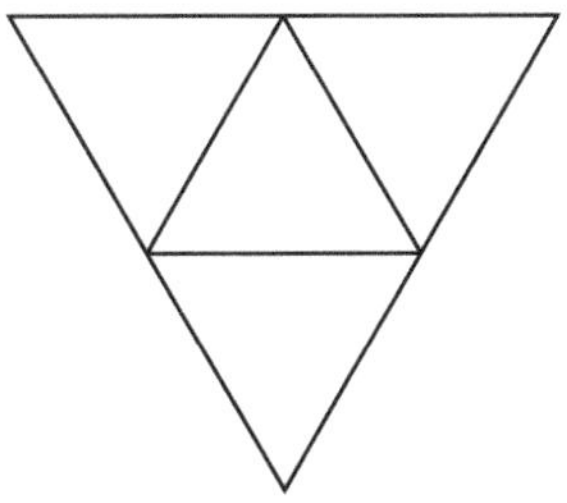

____

c 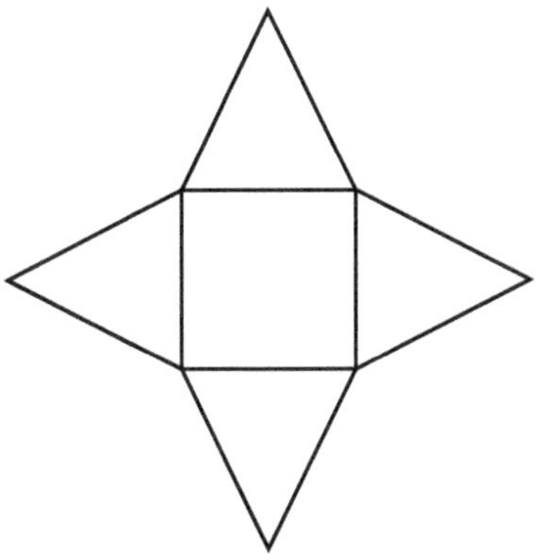

____

d 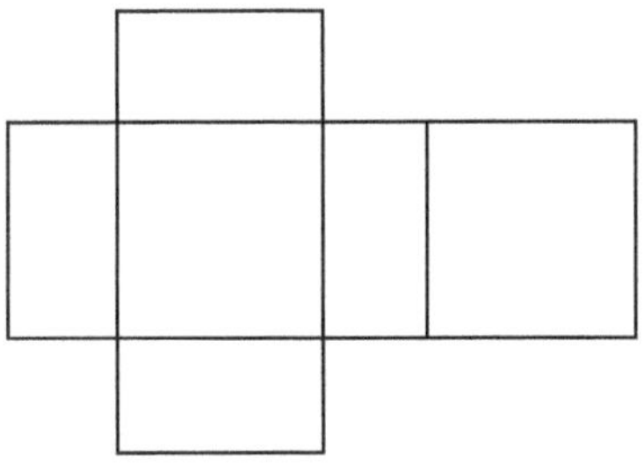

____

65. a You will need squared paper. Use a ruler to create a shape with at least 2 reflex angles, 2 acute angles and 2 obtuse angles. Your lines can overlap. Write on your diagram what type the angles are and label them A, B, C, D, E and F.

b Estimate the size of the angles you have drawn and fill in the estimates in the table. Using a protractor, measure the angles you have drawn and complete the table.

Name of angle	Estimate	Actual measurement
A		
B		
C		
D		
E		
F		

66. Calculate the missing complementary and supplementary angles.

a

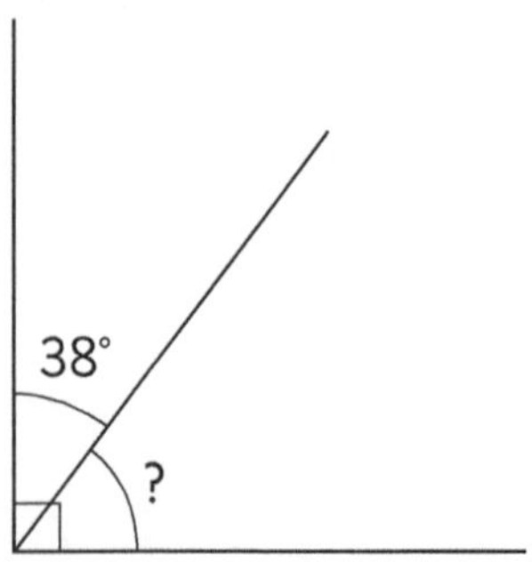

c

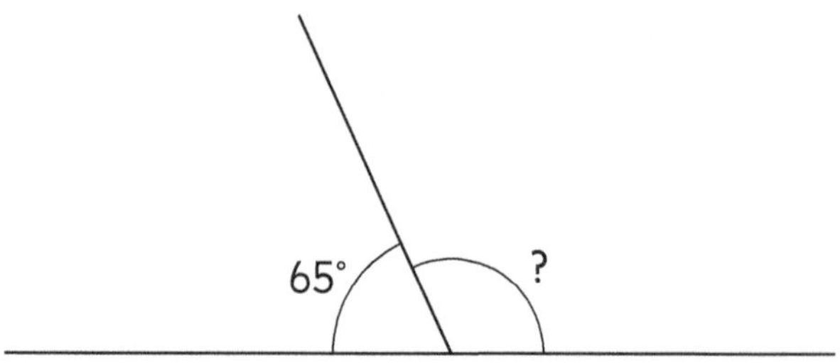

b

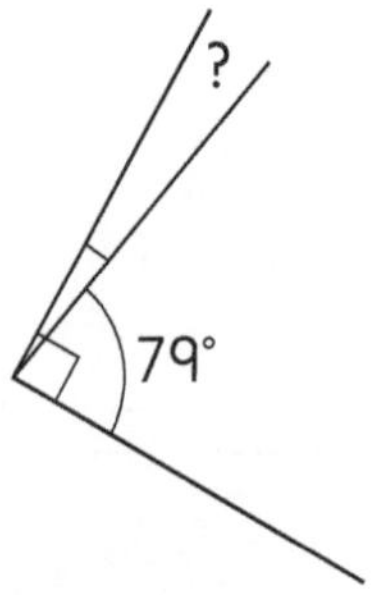

d

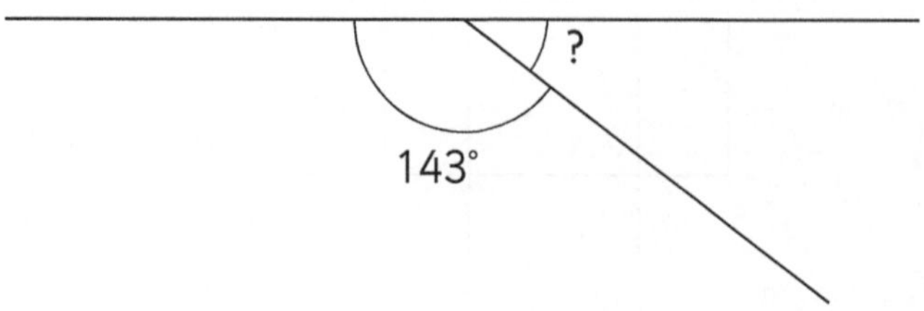

67. How many degrees lie between:

    a north and south? ________________

    b north and north-east? ________________

    c north and west ________________

    d north and south-west? ________________

68. a Write down the coordinates of the vertices of this shape, starting at (1, 8) and then continuing clockwise.

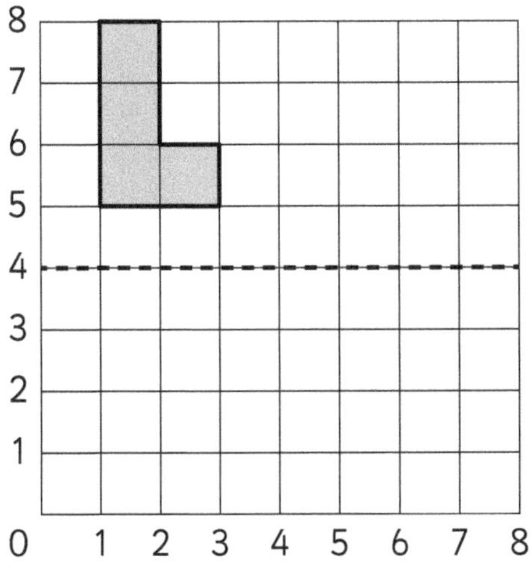

____

b The L-shape is reflected in the dotted line of symmetry. Draw the reflected shape. Then, starting at (1, 0), write down the coordinates of the vertices of the reflection.

____

69. Which of these letters of the alphabet have two lines of symmetry? Draw the two lines of symmetry onto the letters.

A **B** **H** **I** **X**

70. Complete the design for each square tile by reflecting the square pattern in both lines of symmetry.

a

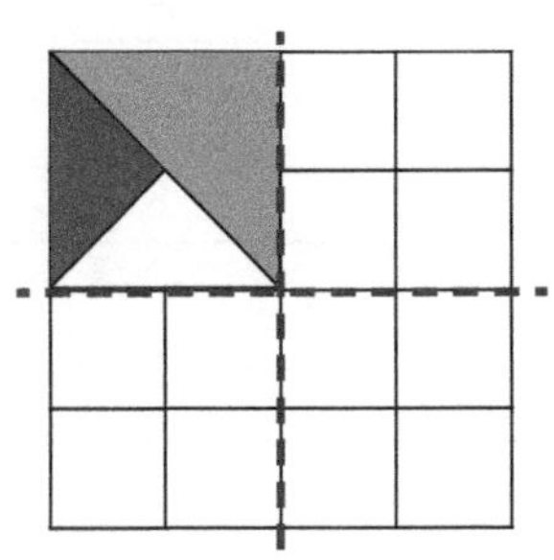

b

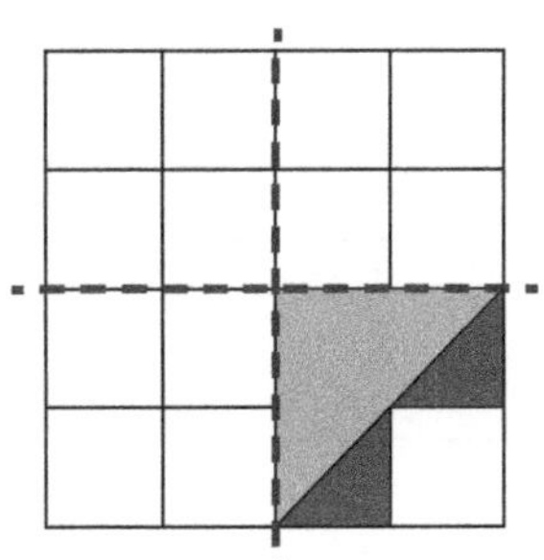

Yearly progress check 2B

71. This is the map of a local town. Point A is the hospital, B is the school, C is the leisure centre and D is the cinema.

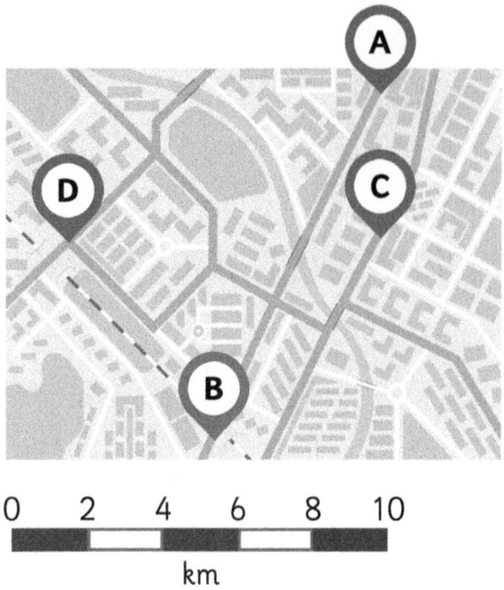

a What is the scale of this map?

____

b How many km is it from the school to the hospital?

____

c How far is it from the cinema to the leisure centre?

____

d Approximately how far will it be from the school to the leisure centre? How do you know?

____

____

____

____

____

Yearly progress check 2B

72. **Number of insects observed from January to July**

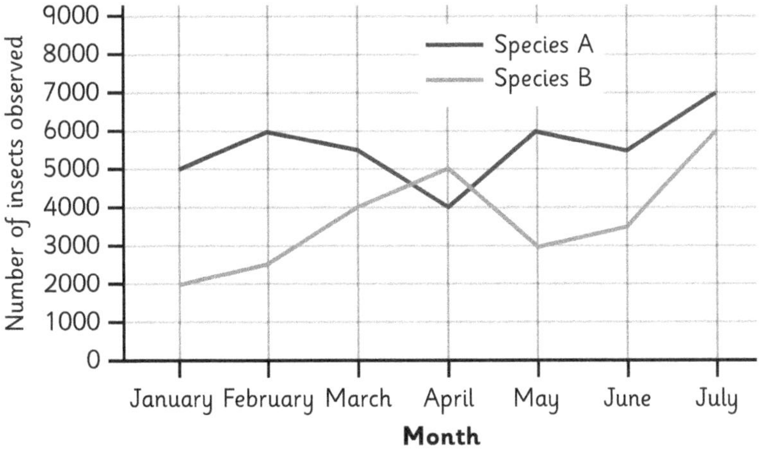

a In January, how many more of Species A than Species B were counted?

____

b Which months showed the least difference between species numbers? How much was this?

____

73. **Holiday destinations**

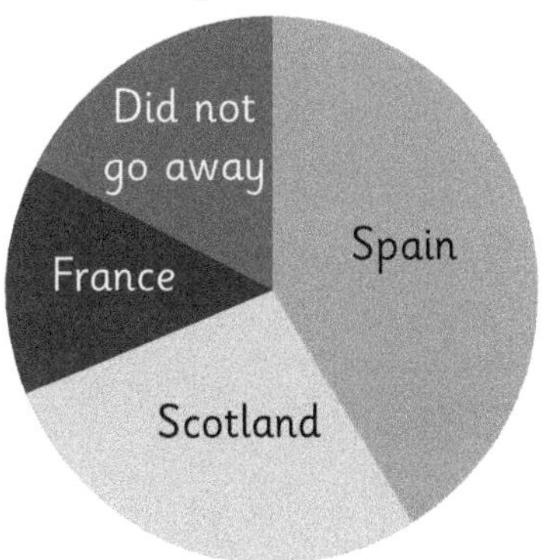

360 people were asked where they went on their summer holiday. Measure each section carefully using a protractor to answer the questions.

a Which was the least popular holiday destination?

____

b How many people went to Spain on holiday?

____

c How many more people went to somewhere in Scotland than did not go away?

____

74. What conclusions about the reliability of data can we draw from these statements. Justify your answers.

    a 'Dogs make the best pets!' – a quote from 'Dogs Today' magazine.

    b 'The children of Scotland's favourite hobby is playing computer games.' – from a survey in P6 Parkmill Primary School.

    c 'You should be aiming to eat at least five pieces of fruit or vegetables every day.' – recommendation from the National Health Service.

    d 'Fizzy drinks are not bad for your health.' – quote from the Head of Fizzy Drinks company.

75. Write a statement for each word on the probability scale (impossible, unlikely, even chance, likely, certain) that is true for you.

Impossible	***Unlikely***	***Even Chance***	***Likely***	***Certain***
0	***0·25***	***0·5***	***0·75***	***1***
0%	***25%***	***50%***	***75%***	***100%***
0	$\frac{1}{4}$	$\frac{2}{4}$ ***or*** $\frac{1}{2}$	$\frac{3}{4}$	***1***

<table>
<tr><th>Topic – benchmarks / Es & Os</th><th>Question</th><th>Answer</th><th>Marking guidance</th></tr>
<tr><td>Estimation and rounding
MNU 2-01a</td><td>1</td><td>3756 rounded to the nearest 100 is 3800.

3756 rounded to the nearest 1000 is 4000.</td><td>On track<ul><li>Understands the conventions of rounding – rounding up when the ones digit is 5 or more, rounding down when the digit is 4 or less.</li><li>Can round to the nearest 100 and 1000 accurately.</li></ul>Further learning required<ul><li>May use incorrect rounding strategies, e.g. focusing on the incorrect digit for rounding; rounding to the nearest thousand instead of the nearest hundred.</li></ul></td></tr>
<tr><td>MNU 2-01a</td><td>2</td><td><table><tr><th></th><th>Nearest tenth</th><th>Nearest whole number</th></tr><tr><td>4.56</td><td>4.6</td><td>5</td></tr><tr><td>101.01</td><td>101.0</td><td>101</td></tr><tr><td>150.52</td><td>150.5</td><td>151</td></tr></table></td><td>On track<ul><li>Can round decimal fractions accurately to the nearest tenth or whole number.</li><li>Understands the conventions of rounding – rounding up when the digit is 5 or more, rounding down when the digit is 4 or less.</li></ul>Further learning required<ul><li>May confuse when to round up or round down, e.g. uses all 52 hundredths (not just the 2 hundredths) to round 150.52 up to 150.6 when rounding to the nearest tenth.</li><li>May believe all decimal fractions must be rounded to the nearest whole number.</li></ul></td></tr>
<tr><td>MNU 2-01a</td><td>3</td><td>4829 + 3755 = 8584
Finlay's estimate is closest to the actual answer.
Finlay rounds to the nearest hundred so his calculation will be 4800 + 3800 = 8600. Nuria is rounding to the nearest thousand, meaning her estimate of 5000 + 4000 = 9000 will not be as close to the actual answer.</td><td>On track<ul><li>Understands the conventions of rounding and can explain why rounding to the nearest hundred will give a more accurate answer in this example.</li></ul>Further learning required<ul><li>May be unclear about how to round to the nearest hundred or thousand</li><li>May not appreciate the impact of an inappropriate choice on the accuracy of answers.</li></ul></td></tr>
</table>

Topic – benchmarks / Es & Os	Question	Answer	Marking guidance
Number – order and place value MNU 2-02a	4	**Amman is incorrect thinking that 80 080 is eight hundred and eighty.** He is saying the numbers he sees and not thinking about the value of the digits or using zero as a place holder. **The number is eighty thousand and eighty.**	**On track** • Understands 'zero' as a place holder in five-digit numbers and can explain Amman's mistake clearly. • Writes number words accurately. Although accurate spelling should be encouraged, reasonable and recognisable attempts (e.g. thowsind) should be accepted). **Further learning required** • May be confused about the 'zero' place holder and so is unable to explain why Amman is incorrect. • May write the five-digit number incorrectly, e.g. eighty hundred eighty.
MNU 2-02a	5	Isla – **67 035** Amman – **32 320**	**On track** • Understands the relationship between the place value arrow cards/counters and the digits that they represent in a 5-digit number. **Further learning required** • May be confused about where to position the place holder '0'. • May write five-digit numbers incorrectly, e.g. by confusing thousands for tens of thousands or omitting the '0'.
MNU 2-02a	6	a 2130 can be made with: 21 × 100 and 3 × 10 OR 21 × 100 and 30 × 1 OR 213 × 10 OR 2130 × 1 b 1004 can be made with: 10 × 100 and 4 × 1 OR 100 × 10 and 4 × 1 OR 1004 × 1 Other combinations of 100s, 10s and 1s are possible.	**On track** • Flexibly partitions 2130 and 1004 using only hundreds, tens and ones. **Further learning required** • May be unable to 'see' 2130 as anything other than 2 thousands, 1 hundred and 3 tens.
MNU 2-02a	7	a 76 539, 77 539, **78 539, 79 539,** 80 539, 81 539, 82 539, **83 539, 84 539** b 10 560, 10 550, 10 540, **10 530, 10 520,** 10 510, **10 500, 10 490,** 10 480 c 20 350, 20 250, 20 150, **20 050, 19 950, 19 850,** 19 750, 19 650	**On track** • Can complete each sequence by increasing or decreasing in increments of 10, 100 or 1000. **Further learning required** • May be unable to complete sequences that bridge a ten, hundred or thousand, e.g. 79 539 to 80 539, 10 500 to 10 490 or 20 150 to 20 050 to 19 950.

Topic – benchmarks / Es & Os	Question	Answer	Marking guidance
MNU 2-02a	8	a 41 649 < 43 649 b 27 270 > 27 027 c 48 020 < 48 200	**On track** • Can correctly use the < and > symbols to create accurate statements. **Further learning required** • May confuse the < and > signs.
MNU 2-04a	9	a −16 b −13 c −10 d −4 e 0 f 5 g 12 h 19 Temperatures in order from coldest to hottest: −49°C, −12°C, 0°C, 3°C, 15°C Coldest temperature: minus forty-nine degrees Celsius Hottest temperature: fifteen degrees Celsius	**On track** • Can correctly place positive and negative numbers on a number line. • Can write temperatures above and below zero degrees in numbers and words. Although accurate spelling should be encouraged, reasonable and recognisable attempts (e.g. ninteen) should be accepted. • Can order temperatures below and above zero, from coldest to hottest. **Further learning required** • May not understand how to write or position negative numbers. • May be unable to write temperatures below zero accurately, e.g. may use the word negative. • May confuse numbers/ temperatures below zero with those above zero so may order the given temperatures using the numbers alone e.g. 0, 3, −12, 15, −49.
MNU 2-02b	10	a **0.47**; 47 small squares coloured on the hundred square– **zero point four seven** is written b **1.25**: 1 large square plus 25 small squares coloured on the hundred squares – **one point two five** is written c **2.03**: two whole squares plus 3 small squares coloured on the hundred squares – **two point zero three** is written d $5.46 = 5\frac{46}{100}$ e $3.03 = 3\frac{3}{100}$ f $12.98 = 12\frac{98}{100}$	**On track** • Decimal fractions are accurately represented on the hundred squares and written correctly in words. Although accurate spelling should be encouraged, reasonable and recognisable attempts should be accepted. For example, 'one point twenty-five' is incorrect but an answer of 'one pint tow five' should be accepted (questions a,b,c). • Decimal fractions are written correctly as mixed numbers (questions d,e,f). **Further learning required** • May be able to represent 0.47 correctly on squared paper but have difficulty representing decimal numbers greater than one.

Topic – benchmarks / Es & Os	Question	Answer	Marking guidance
			• May be unable to write decimal fractions accurately in words, e.g. 0.47 may be written as zero point forty-seven. • May struggle to convert decimal fractions into mixed numbers e.g. 3.03 may be written as $3\ \frac{3}{10}$
MNU 2-03b	11	Can make three of the following decimal fractions: **2.06, 2.09, 2.36, 2.39, 2.76, 2.79** **6.06, 6.09, 6.36, 6.39, 6.76, 6.79**	**On track** • Can select place value arrow cards to make three different numbers each with 2 decimal places. • May notice that it is possible to represent decimal numbers with 3 digits, where one digit is a zero, using only two place value arrow cards. **Further learning required** • May be able to create decimal fractions with place value cards but have difficulty recording their answers or make errors, e.g. may write 2.9 instead of 2.09
MNU 2-20a	12	a 3.72 b 3.79 c 3.85 d 3.92 e Smallest to largest: 2.8, 2.85, 5, 5.28, 5.82, 28	**On track** • Can accurately identify the decimal fraction represented by each arrow on the number line. • Can order ragged decimal fractions (decimal fractions with differing number of digits after the decimal point) from smallest to largest. **Further learning required** • May not be able to work out that each graduation on the number line represents one hundredth and so confuse the decimal fractions, e.g. may write 3.8 or 3.9 instead of 3.72. • May ignore the decimal point and so believe 2.8 and 28 are equal and/or answer 5, 2.8, 28, 2.85, 5.28, 5.82. • May believe the fewer digits the number has the smaller its value and vice-versa, e.g. thinks 5 is the smallest number and 5.82 the largest.

Topic – benchmarks / Es & Os	Question	Answer	Marking guidance
Number – addition and subtraction MNU 2-02a MNU 2-03a	13	a **False** – 2853 is one more than 2852 and 1375 is one more than 1374, so the two calculations cannot be equal. b **True** – there is a difference of one between 7999 and 8000, and 3999 and 4000. As both are addition calculations, this means the answers will be equal. c **True** –addition is commutative so the numbers can be added in any order. d **False** – the starting number in the second calculation is 9 more than in the first. However, the number you are subtracting is nine less in the second calculation. This means the two calculations cannot be equal.	**On track** • Calculates accurately and can explain why each statement is true or false, using informal jottings to check their answers. • Is able to clearly explain how answers have been worked out. May use strategies such as compensation (mentally), jumps on an empty number line, or informal jottings. May also use different strategies depending on the calculation. **Further learning required** • May confuse the properties of addition and subtraction or may need to work out each calculation individually in order to answer 'true' or 'false'.
MNU 2-03a	14	a 229 + 111 + 1238 = **1578** b 3000 + 2590 + 410 = **6000** c 8000 − 4760 − 2240 = **1000** * links should be made to making 10s, 100s or 1000s to support calculations.	**On track** • Can use multiples of 10, 100 and 1000 to simplify calculations, e.g. in question a changes the calculation to 230 + 110 + 1238 = 340 + 1238 **Further learning required** • May be unable to round to the nearest 10, 100 or 1000 mentally. • May have difficulty identifying numbers that total a multiple of 1000, e.g. may not realise that 2240 can be added to 4760 then subtracted from 8000 in question c. • May rely on formal, written methods.
MNU 2-02a MNU 2-03a	15	a 89 119 = **3000** + 86 119 b 54 700 = **75 700** − 21 000	**On track** • Can use knowledge of place value to complete each number sentence accurately. • Understands how to complete an equation using inverse relationships. **Further learning required** • Does not understand how to complete an equation. • May not recognise how to use the inverse relationship so wrongly adds 89 119 and 86 119 or subtracts 21 000 from 54 700.

Topic – benchmarks / Es & Os	Question	Answer	Marking guidance
MNU 2-02a MNU 2-03a	16	a **5728 + 4258 =9986** 5000 + 4000 = 9000 700 + 200 = 900 20 + 50 = 70 8 + 8 = 16 b **91 766 – 81 510 =10 256** 91 000 – 81 000 = 10 000 700 – 500 = 200 60 – 10 = 50 6 – 0 = 6 OR 66 – 10 = 56	**On track** • Can add and subtract four and five-digit numbers by partitioning them into thousands, hundreds tens and ones. • Partitions mentally and records each step in the calculation using informal jottings or a semi-formal written method. **Further learning required** • May be unable to partition mentally and require the support of base-10 material. • May be able to use place value partitioning to add two numbers but have difficulty with subtractions. Notably, he/she may not understand that the result of each step must be **added** to find the answer.
MNU 2-03a	17	a 5786 + 2**931** = **8**71**7** b 4755 + **2689** = 7444	**On track** • Can use knowledge of number bonds, place value and the conventions of the standard written method to complete each calculation. • The pupil may use a checking strategy, e.g. by recalculating or using an estimated 'sense check' of the size of the numbers. Evidence of this may not be written and you will need to ask if this is a strategy they are using. **Further learning required** • May calculate incorrectly, especially where carrying is required e.g. may answer 2789 for b. • May attempt to use the inverse relationship but does so incorrectly.
MNU 2-03a	18	a The tens row should **read –30**, making the answer **371**. b The tens row should read **10** not –10. The correct answer is therefore **404**.	**On track** • Can use semi-formal methods to subtract three-digit numbers from three-digit numbers. **Further learning required** • May be unable to use place value partitioning to help support their calculations. • May be confused by examples where the subtraction results in a negative number, e.g. may think they can subtract 60 from 70 instead of 70 from 60 and/or add rather than subtract negative numbers.

Topic – benchmarks / Es & Os	Question	Answer	Marking guidance
MNU 2-03a	19	Nuria's family still have **£388** to save for their holiday.	**On track** • Understands the word problem, i.e. identifies subtraction as the required operation. • Can partition numbers flexibly and solve the problem using a strategy of their choice, e.g. by counting on or back on an empty number line or by using a semi-formal or formal written method. **Further learning required** • May be unable to interpret and/or visualise the word problem and so chooses the incorrect operation, i.e. adds 867 and 479. • May be unable to choose an effective strategy and/or make simple calculation errors.
MNU 2-03a	20	Answer will vary according to the year. For example, in the year 2020, Caroline Herschel discovered the first comet **237** years ago.	**On track** • Appreciates the problem involves finding the difference between the current year and 1783. • Appreciates that they need to find the difference between two years and can count on or back on an empty number line to solve the problem. • Can write a calculation that mirrors how they solved the problem. **Further learning required** • May be unable to identify the unknown and so have difficulty representing the problem as a bar model. • May be unable to select and use an appropriate strategy to solve the problem. • May make simple calculation errors. • May use ineffective or inefficient strategies when calculating, e.g. counting on or back in 10s and 100s on an empty number line instead of multiples of 10 or 100.

Topic – benchmarks / Es & Os	Question	Answer	Marking guidance
MNU 2-03a	21	2000 tickets were left unsold. 7100 + (2 × 3450) = 14 000 16 000 – 14 000 = 2000	**On track** • Can correctly interpret the problem and identify the calculations required to solve it. • Can choose an effective strategy for calculation, dependant on the numbers and the type of calculation involved; for example, mental strategies supported by informal jottings/diagrams if necessary. • Can flexibly partition numbers in order to calculate. • Can explain how they solved the problem and justify the strategies used. **Further learning required** • May carry out the wrong calculations or calculate inaccurately. • May use ineffective or inefficient strategies when calculating, e.g. counting on or back in 10,100 or 1000 on an empty number line instead of multiples of 10, 100 or 1000. • May be over reliant on one strategy that may not be the most efficient. • May not be able to explain which strategy was used and why.
MNU 2-03b	22	a 169 + 13.7= **182.7** b 51.8 + 20.7 = **72.5** c 70.7 + 2.4 = **73.1** d 31.4 + 25.9 = **57.3**	**On track** • Can choose an effective strategy for calculation, dependant on the numbers and the type of calculation. • Can flexibly partition numbers in order to calculate. • Can explain and justify their choice of strategy. **Further learning required** • May calculate inaccurately. • May rely on using the same strategy for each calculation. • May be unable to mentally bridge a one or a ten.

Topic – benchmarks / Es & Os	Question	Answer	Marking guidance
MNU 2-03b	23	a 24.9 + 38.1 = **63** 38.1 + 24.9 = **63** **63** – 38.1 = 24.9 **63** – 24.9 = 38.1 b **20.5** + 60.5 = 81 60.5 + **20.5** = 81 81 – 60.5 = **20.5** 81 – **20.5** = 60.5 c **45.9** + 54.1 = 100 54.1 + **45.9** = 100 100 – 54.1 = **45.9** 100 – **45.9** = 54.1 d 79.4 + 40.6 = **120** 40.6 + 79.4 = **120** **120** – 79.4 = 40.6 **120** – 40.6 = 79.4	**On track** • Understands the part-whole relationship and can write two addition and two subtraction number sentences to match each bar model. • Can use flexible partitioning to mentally calculate the missing values, supported by an empty number line or informal jottings where necessary. **Further learning required** • May be able to solve problems where the whole is unknown but find examples where one of the parts is unknown more challenging. • May choose the wrong operation, e.g. adds 81 and 60.5 instead of working out what is added to 60.5 to make 81. • May make calculation errors. • May be unable to write two addition and two subtraction number sentences to match each bar model. For example, may only see a as 24.9 + 38.1.
Number – multiplication and division MNU 2-03a	24	a 6 × 2 = **3 × 2 × 2 = 12** b 6 × 5 = **3 × 2 × 5 = 30** c 7 groups of 6 = **7 × 3 × 2 = 42** d 18 divided by 6 = **18 ÷ 3 ÷ 2 = 3** Arrays may be drawn to demonstrate the relationship between the 6 times table and the 3 times table.	**On track** • Understands the concept of doubling and can apply this knowledge to 3 times table facts to multiply by 6. • Can demonstrate the relationship between the 6 times table and the 3 times table; for example, by drawing an array. **Further learning required** • May be unable to recall 3 times table facts. • May be able to draw an array to represent the calculation but may be unable to link this to the 3 times table.
MNU 2-03a	25	Students may explain their recall of the 9 times table by: Using the 10 times table then taking one group away e.g. 3 × 9 = (3 × 10) – (3 × 1) or 10 threes minus 3	**On track** • Can accurately explain at least one strategy to help recall of the 9 times table. **Further learning required** • May be unable to explain a strategy to help support the recall of the 9 times table.

Topic – benchmarks / Es & Os	Question	Answer	Marking guidance
MNU 2-02a MNU 2-03a	26	8500 divided by 10 is 850, so **Amman** is correct. When a number is divided by 10, the value of each digit becomes 10 times smaller.	**On track** • Can divide a multiple of 100 by 10 accurately. • Can explain what happens when a multiple of 100 is divided by 10. **Further learning required** • May be unable to identify who is correct and/or justify their choice. • May be confused between dividing by 10 and 100. • May answer correctly but use an incorrect explanation, e.g. 'When you divide a number by 10 you take a zero away'.
MNU 2-03a	27	a 360 ÷ 6 **Known fact: 36 ÷ 6 = 6** So, 360 ÷ 6 =**60** b 2400 ÷ 4 **Known fact: 24 ÷ 4 = 6** So, 2400 ÷ 4 = **600** c 210 ÷ 30 **Known fact: 21 ÷ 3 = 7** So, 210 ÷ 30 =**7**	**On track** • Can use known facts and knowledge of multiplying and dividing by 10 and 100 to calculate all answers correctly. **Further learning required** • May be unable to recall multiplication and division facts. • May confuse multiplying and dividing by 10 and 100. • May be able to divide 3- and 4-digit multiples of 10 and 100 by a single digit but find dividing these numbers by a multiple of 10 more challenging.
MNU 2-03a	28	a 4 b 8 c 6	**On track** • Can interpret and visualise the problem in terms of 'equal groups'. • Understands division as the inverse of multiplication and uses known facts to solve the problem mentally. **Further learning required** • May be unable to recall multiplication and division facts and rely on forming equal groups of concrete materials to solve the problem. • Does not understand how to solve a grouping problem using division, i.e. is unable to choose the correct operation.

<table>
<tr><th>Topic – benchmarks / Es & Os</th><th>Question</th><th>Answer</th><th>Marking guidance</th></tr>
<tr><td>MNU 2-03a</td><td>29</td><td>a 378×5
$= (300 \times 5) + (70 \times 5) + (8 \times 5)$
$= 1500 + 350 + 40$
$= 1890$

b $96 \div 4$
$= (80 \div 4) + (16 \div 4)$
$= 20 + 4$
$= 24$
Allow reasonable alternative partitioning strategies and layout of their working.</td><td>On track<ul><li>Can use known facts and knowledge of multiplying and dividing.</li><li>Understands how to use standard and non-standard partitioning.</li><li>Can apply the strategy of partitioning to solve multiplication and division problems using brackets to organise their working.</li></ul>Further learning required<ul><li>May be unable to recall multiplication and division facts.</li><li>Cannot use standard and non-standard partitioning.</li><li>Does not choose an appropriate partition, making the problem hard to solve e.g. $96 \div 4 = (90 \div 4) + (6 \div 4)$.</li><li>Assumes division is always related to subtraction, so inserts a minus sign in their calculation, e.g. $96 \div 4 = (80 \div 4) - (16 \div 4)$.</li><li>Misses out some steps or makes calculation errors, e.g. $300 \times 5 = 150$.</li></ul></td></tr>
<tr><td>MNU 2-03a</td><td>30</td><td>a 432×8
$432 = 400 + 30 + 2$
<table><tr><td>×</td><td>400</td><td>30</td><td>2</td></tr><tr><td>8</td><td>3200</td><td>240</td><td>16</td></tr></table>$3200 + 240 + 16 = 3456$

b $78 \div 3$
$78 = 60 + 18$
<table><tr><td>÷</td><td>60</td><td>18</td></tr><tr><td>3</td><td>20</td><td>6</td></tr></table>$20 + 6 = 26$</td><td>On track<ul><li>Can use known facts and knowledge of multiplying and dividing.</li><li>Understands how to use standard and non-standard partitioning.</li><li>Can apply the strategy of partitioning to solve multiplication and division problems using a grid to organise their working.</li></ul>Further learning required<ul><li>May be unfamiliar with the grid method.</li><li>May be unable to recall multiplication and division facts.</li><li>Cannot use standard and non-standard partitioning.</li><li>Does not choose an appropriate partition, making the problem hard to solve, e.g. $78 \div 3 = (70 \div 3) + (8 \div 3)$.</li><li>Assumes division is always related to subtraction, so inserts a minus sign in their calculation, e.g. instead of $20 + 6 = 26$ writes $20 - 6 = 14$.</li><li>Misses out some steps or makes calculation errors, e.g. $18 \div 3 = 9$.</li></ul></td></tr>
</table>

Topic – benchmarks / Es & Os	Question	Answer	Marking guidance
MNU 2-03a	31	6×29 can be worked out by multiplying by 30. 6×30 is 180. Then subtract 6×1. $180 - 6 =$ **174**	**On track** • Knows what a multiple of 10 is. • Understands the relationship between 6×3 and 6×30. • Knows how to calculate using a compensation strategy. **Further learning required** • May not understand how 6×3 relates to 6×30. • May add on 6 instead of subtracting 6. • May not know what a multiple is.
MNU 2-03a	32	a $6 \times 6 = 36$ Amman picked 36 punnets b $8 \times 6 = 48$ $48 - 36 = 12$ Nuria picked 12 more punnets than Amman.	**On track** • Can visualise the word problem and make links between that and the multiplication calculations. • Identifies multiplication as the appropriate operation and can use their knowledge of the 6 times table to find how many punnets each character picked. • Can solve multi-step problems, i.e. knows they must first calculate the number of punnets picked by Nuria and find the difference between the two amounts. **Further learning required** • May be unable to represent or visualise the word problem and/or carry out the appropriate calculations. • May omit a step when calculating, e.g. may answer 48 instead of 12 in part b.

Topic – benchmarks / Es & Os	Question	Answer	Marking guidance
MNU 2-02a MNU 2-03a	33	a $4.7 \times 10 = \mathbf{47}$ b $15.2 \times \mathbf{10} = 152$ c $3.4 \times \mathbf{1000} = 3400$ d $\mathbf{4600} \div 10 = 460$ e $427 \div 10 = \mathbf{42.7}$ f $9100 \div 1000 = \mathbf{9.1}$	**On track** • Can use a place value grid (if required) to complete calculations involving decimal fractions. • Can confidently multiply and divide whole numbers and decimal fractions with one decimal place by 10, 100 and 1000 mentally. • Can use their knowledge of the inverse relationship between multiplication and division to solve non-standard problems, i.e. b, c and d. **Further learning required** • May be unable to use place value to complete the calculations, even with the support of a place value grid. • May not understand the inverse relationship between multiplication and division. • May confuse multiplying and dividing. • May be able to multiply and divide whole numbers by 10, 100 and 1000 but have difficulty multiplying and dividing decimal numbers by 10, 100 and 1000. This may result from picking up unhelpful and mathematically incorrect 'rules' about adding and subtracting zeros.
MNU 2-03b	34	a Each of the four friends should pay **£9.50**. b Nuria ran **half a kilometre/0.5 km** each day.	**On track** • Can visualise and/or represent each problem using concrete materials, pictures or diagrams and make links with multiplication/division calculations. • Can accurately complete calculations involving dividing whole numbers and decimals. **Further learning required** • May be unable to visualise or represent the word problems, choose the wrong operation or make calculation errors. • May not understand how to split wholes into smaller parts (halves) e.g. for question a may know that $4 \times 9 = 36$ but may not be sure how to divide the extra £2 between the four friends.

Topic – benchmarks / Es & Os	Question	Answer	Marking guidance
Multiples, factors and primes MNU 2-05a	35	a Factors of 8: 1, 2, 4 and 8 b Factors of 12: 1, 2, 3, 4, 6 and 12 c Factors of 18: 1, 2, 3, 6, 9 and 12	**On track** • Can identify factors of given numbers. • Can list all factors of given numbers. **Further learning required** • May not understand what 'factors' means. • May identify some but not all factors of the given numbers.
MNU 2-05a	36	6 **8** **23** 27 12 **17** 42 60 8, 23 and 17 are not multiples of 3. These numbers do not divide exactly by three (or are not products of 3/'answers in the 3 times table').	**On track** • Can understand the vocabulary of 'multiples'. • Can explain why certain numbers are multiples of 3 and others are not. **Further learning required** • May not understand what 'multiples' means. • May be confused regarding which numbers are multiples of three and which are not, e.g. may think 23 is a multiple of 3 because the digit '3' is present.
Fractions, decimal fractions and percentages MNU 2-07a	37	Halves and fifths are the only fractions that can equal tenths.	**On track** • Can show fractions equivalent to tenths on the fraction wall, so 1 half = 5 tenths; 1 fifth = 2 tenths; 2 fifths = 4 tenths etc. **Further learning required** • Can name fractions on the fraction wall but is unable to convert, for example, fifths into tenths. • May only be able to identify unit fractions equivalent to tenths, i.e. one half = five tenths, one fifth = two tenths.
MNU 2-07b	38	a $\frac{2}{3} = \frac{4}{6}$ and $\frac{8}{12}$ b $\frac{3}{5} = \frac{6}{10}$ c $\frac{1}{2} = \frac{2}{4}, \frac{3}{6}, \frac{4}{8}, \frac{5}{10}$, and $\frac{6}{12}$	**On track** • Can write equivalent fractions using their fraction wall. **Further learning required** • May be unable to read the fractions given correctly, e.g. reads $\frac{2}{3}$ as two quarters. • May be unable to equate the fractions given with the fraction wall. • May be unable to find equivalent fractions or writes them incorrectly.

Topic – benchmarks / Es & Os	Question	Answer	Marking guidance
MNU 2-07b MNU 2-07c	39	a $\frac{1}{4}\left(\frac{5}{20}\right)$, $\frac{2}{5}\left(\frac{8}{20}\right)$, $\frac{1}{2}\left(\frac{10}{20}\right)$, $\frac{6}{10}\left(\frac{12}{20}\right)$, $\frac{14}{20}$ placed in the correct position on the number line. b **Finlay eats more cake.** Isla had $\frac{2}{3}$ of the cake left. She cut this into six slices, meaning that each third is now in three pieces. The whole cake would have been $\frac{9}{9}$. Isla has eaten $\frac{3}{9} + \frac{2}{9} = \frac{5}{9}$. 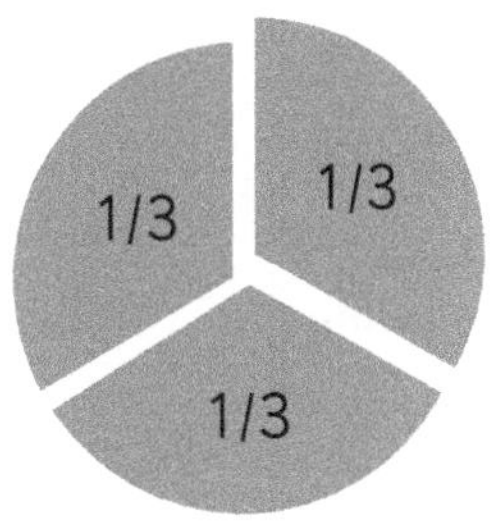 Finlay had $\frac{1}{2}$ the cake left. He cut this into four slices, meaning that the remaining half is now in four pieces. The whole cake would have been $\frac{8}{8}$. Finlay has eaten $\frac{4}{8} + \frac{1}{8} = \frac{5}{8}$. 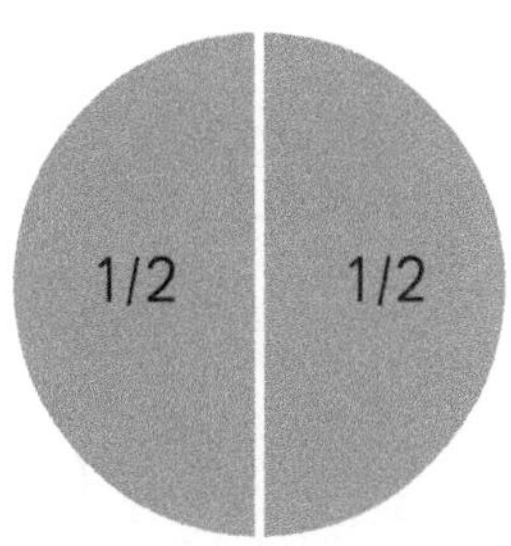 Ninths are smaller than eighths when comparing the same-sized whole, so $\frac{5}{8}$ is more than $\frac{5}{9}$.	**a On track** • Can represent and solve the problem, drawing diagrams or using concrete materials to help them visualise the fractions involved, if necessary. • Understands that twenty equal sections on the number line means that each section represents $\frac{1}{20}$. • Can make links with multiplication and division to work out equivalent fractions. • Can use some or all of the above to order the fractions from smallest to largest. **Further learning required** • May be unable to create equivalent fractions beyond $\frac{1}{2} = \frac{10}{20}$, with concrete materials or pictorially. • May base their decision about the order purely on the denominators, i.e. thinks $\frac{1}{2}$ is the smallest fraction because 2 is smaller than 4, 5, 10, 20. Conversely, thinks $\frac{14}{20}$ is the smallest fraction because twentieths are smaller than halves. • May base their decision about the order purely on the numerators, i.e. thinks the larger the numerator the larger the fraction. **b On track** • Can represent and solve the problem by drawing diagrams to help them visualise the fractions involved. **Further learning required** • May answer correctly without truly understanding the problem. For example, may have difficulty interpreting the problem so disregard some of the information and decide the answer is Finlay because $\frac{1}{2}$ of the same sized whole will be larger than $\frac{1}{3}$. It is therefore important that pupils are given the opportunity to show and/or explain their thinking.

<table>
<tr><th>Topic – benchmarks / Es & Os</th><th>Question</th><th>Answer</th><th>Marking guidance</th></tr>
<tr><td></td><td></td><td></td><td>
<ul>
<li>May be unable to visualise the word problem and/or represent it pictorially.</li>
<li>May confuse Isla cutting $\frac{2}{3}$ of the cake into six pieces, as the cake now being in sixths. This may be because they believe wholes do not have to be the same size in order to compare them easily.</li>
<li>May try to simply add the fraction of cake eaten as whole numbers e.g. $\frac{1}{3} + \frac{2}{6} = \frac{3}{9}$</li>
</ul>
</td></tr>
<tr><td>MNU 2-07a
MNU 2-07b</td><td>40</td><td>
<table>
<tr><th></th><th>Whole and hundredths e.g. 1 whole and eighteen hundredths</th><th>Hundredths e.g. 118 hundredths</th><th>Decimal fraction e.g. 1.18</th></tr>
<tr><td>1.27</td><td>1 whole and 27 hundredths</td><td>127 hundredths</td><td>1.27</td></tr>
<tr><td>2.56</td><td>2 wholes and 56 hundredths</td><td>256 hundredths</td><td>2.56</td></tr>
<tr><td>3</td><td>3 wholes</td><td>300 hundredths</td><td>3 (accept 3.0 or 3.00)</td></tr>
</table>
</td><td>
On track
<ul>
<li>Can identify the fraction represented by each picture.</li>
<li>Can write this fraction as wholes and hundredths, hundredths and as a decimal or whole number.</li>
</ul>
Further learning required
<ul>
<li>May be unable to equate the picture with the correct fraction.</li>
<li>May be unable to write the fraction in different ways, e.g. may write 1 whole and 27 hundredths but struggle to write this as hundredths only and/or as a decimal number.</li>
</ul>
</td></tr>
<tr><td>MNU 2-07b
MNU 2-07c</td><td>41</td><td>
One quarter, three quarters and three fifths can be converted into hundredths.
$\frac{3}{4} = \frac{75}{100} = 0.75$
$\frac{3}{5} = \frac{6}{10} = \frac{6}{100} = 0.6$ (accept 0.60)
$\frac{1}{4} = \frac{25}{100} = 0.25$
$\frac{1}{7}$ cannot be converted into hundredths
</td><td>
On track
<ul>
<li>Can identify which fractions can be converted into tenths and/or hundredths.</li>
<li>Can convert these fractions into tenths and/or hundredths accurately.</li>
</ul>
Further learning required
<ul>
<li>May be unable to identify which fractions can be converted into tenths and/or hundredths.</li>
<li>May be unable to equate the picture with the correct fraction.</li>
<li>May be unable to convert the fractions into tenths and/or hundredths accurately, e.g. may write $\frac{3}{5} = \frac{8}{10}$ (adding five to the denominator and numerator).</li>
</ul>
</td></tr>
</table>

Topic – benchmarks / Es & Os	Question	Answer	Marking guidance
MNU 2-07a	42	a $\frac{3}{4}$ of 76 = **57** b $\frac{5}{6}$ of 132 = **110** c $\frac{7}{10}$ of 5650 = **3955** d $\frac{5}{8}$ = 560. Therefore, to find $\frac{1}{8}$ we calculate 560 ÷ 5 (not divide by 8) = 112 If $\frac{1}{8}$ = 112 then the total population will be $\frac{8}{8}$. 8 × 112 = **896**	**On track** • Can understand and visualise the problems using the bar models. • Can use mental maths strategies and informal jottings to calculate fractions of an amount. • Can interpret the information given in the bar model and choose appropriate strategies to solve it, e.g. known facts, empty number line, jottings (question d). **Further learning required** • May be unable to calculate fractions of an amount or leave out parts of the calculation, e.g. may calculate $\frac{5}{6}$ by dividing by 6 but not multiplying this by 5. • May be unable to interpret, represent and/or visualise the word problem or be unable to choose suitable strategies to solve it.
MNU 2-07a	43	1st – Isla in 35.05 secs 2nd – Nuria in 35.50 secs 3rd – Amman in 37.16 secs 4th – Finlay in 38.20 secs A time between 35.06 secs and 35.49 secs inclusive would mean you would come in 2nd place.	**On track** • Can understand the format of the times to order them from fastest to slowest. • Can identify a time that would beat Nuria, to take second place. **Further learning required** • May equate the largest number with the fastest time, e.g. have Finlay 1st because 38.20 is the largest number. • May be unable to identify a time between 35:50 and 35:05 in order to place themselves second.

Topic – benchmarks / Es & Os	Question	Answer	Marking guidance
MNU 2-07b	44	a 80%, 80 out of 100, $\frac{80}{100}\left(\frac{4}{5}\right)$ b 37%, 37 out of 100, $\frac{37}{100}$ c 55%, 55 out of 100, $\frac{55}{100}\left(\frac{11}{20}\right)$ d 12%, 12 out of 100, $\frac{12}{100}\left(\frac{3}{25}\right)$	**On track** • Can accurately count the shaded boxes OR the number of unshaded boxes to work out the number of seats sold. Counting in groups of 10 would be an efficient strategy to choose. • Can convert the number of boxes shaded to a percentage and a fraction. **Further learning required** • May make counting errors. This may be from counting boxes individually, instead of in groups of 10. • May be able to state that there are 80 out of 100 seats but not equate this with 80% or $\frac{80}{100}$.
MNU 2-07b	45	a $\frac{1}{2} = 50\%$ b $\frac{7}{10} = 70\%$ c $\frac{1}{5} = 20\%$ d $\frac{3}{4} = 75\%$	**On track** • Can accurately shade the hundred square to show the equivalent percentage. • Can write each fraction as a percentage using the percentage symbol. **Further learning required** • May be unable to replicate the original fraction on the hundred square • May be unable to write the equivalent percentage beyond the most common fractions, e.g. $\frac{1}{2}$ or $\frac{3}{4}$.
MNU 2-07a	46	a 10% of 120 is 12. So 60% will be $12 \times 6 = 72$ The farmer has **72 cows**. b A word problem should be written either involving 80% in the calculation and 20% as the answer, or 20% in the calculation and 80% as the answer.	**On track** • Can use mental strategies and/or informal jottings to calculate simple percentages. • Can use knowledge of fractions, multiplying and dividing to support calculations. • Can write a word problem to match a given bar model. **Further learning required** • May be unable to calculate simple percentages or does so incorrectly, e.g. divides 120 by 6 and multiplies by 10 to find 60% of 120. • May be unable to write a word problem that involves 80%/20%.

Topic – benchmarks / Es & Os	Question	Answer	Marking guidance
Money MNU 2-09a	47	a Amman has £17.50. b £5.75 × 3 = £17.25. Yes, Amman does have enough to pay for them all. c No, Amman doesn't have enough money left for popcorn. He only has 25p/£0.25 left, so would need an extra 50p/£0.50 to buy the 75p popcorn.	**On track** • Can visualise and represent the problem, for example using coins and notes or on a number line. • Can identify the calculations required to solve each part of the problem. • Can choose appropriate strategies, supported by informal jottings if necessary, to add, multiply and subtract mentally. **Further learning required** • May be unable to visualise and/or represent the problem and so cannot identify the operation required to solve each part. • May choose an inappropriate strategy, use a strategy incorrectly or make calculation errors. • May guess and be unable to justify their answers to parts b and c.
MNU 2-09a	48	The most Isla will pay per lesson is £5.50 (individual lesson is £4.50 + £1 locker hire)	**On track** • Can visualise, represent and solve the problem, justifying their reasoning. • Can use mental maths strategies, supported by informal jottings if necessary, to add, multiply and compare. **Further learning required** • May not understand the problem and so be unable to identify the calculations required to solve it. • May make calculation errors, e.g. may be unable to compare costs due to inaccuracies with division or misinterpreting information about locker hire.

Topic – benchmarks / Es & Os	Question	Answer	Marking guidance
MNU 2-09c	49	a £4 (or 400p) ÷ 20 = **20p/£0.20 per tub**. b 25p × 20 = £5 or 500p. £5 – £4 = **£1 profit** c 10% of £4 is 40p or £0.40 £4.00 – £0.40 = £3.60 £5 – £3.60 = **£1.40 profit.**	**On track** • Understands the vocabulary 'profit' and 'loss'. • Can visualise and represent each part of the problem; for example, using coins and notes or on a number line. • Can identify the calculations required to solve each part of the problem. • Can choose appropriate strategies, supported by informal jottings if necessary, to add, multiply, subtract and compare. • Can partition numbers effectively to make calculations easier, e.g. multiplying by 20 is calculated either by multiplying by 2 then 10 or multiplying by 10 then 2. **Further learning required** • May be unable to visualise and/or represent the problem and so cannot identify the operation required to solve some or all of the parts. • May be unable to accurately calculate answers, e.g. may confuse working in £ and answering in p. • May omit important steps in calculations, e.g. may calculate the discount the class received but forget to subtract this amount from £5 to show the new profit.

<table>
<tr><th>Topic – benchmarks / Es & Os</th><th>Question</th><th>Answer</th><th>Marking guidance</th></tr>
<tr><td>MNU 2-09b</td><td>50</td><td>a Possible explanations:
• Banks give you debit cards when you open a bank account; credit cards have to be applied for specially.
• Money is taken from your bank account when you use a debit card; a credit card allows you to borrow money.
• You can usually only spend what is in your bank account with a debit card; a credit card allows you to spend up to your spending limit.
b Answers may include:
Advantages:
• Able to spend money that isn't in your bank account.
• You have a PIN number so only you can use the card.
• It can help you pay for expensive goods.
Disadvantages:
• Can get yourself into debt.
• You can spend money you don't have but might not be able to afford the repayments.</td><td>On track
• Can explain the key differences between credit and debit cards.
• Can list potential financial benefits and risks of using credit cards.
Further learning required
• May be unclear about the difference between credit cards and debit cards.
• May be unsure about potential financial benefits and risks of using credit cards.</td></tr>
<tr><td>Time
MNU 2-10a</td><td>51</td><td><table>
<tr><th>12-hour clock</th><th>24-hour clock</th></tr>
<tr><td>10:00 a.m.</td><td>1000</td></tr>
<tr><td>12:00 noon</td><td>1200</td></tr>
<tr><td>4:30 a.m.</td><td>0430</td></tr>
<tr><td>5:45 p.m.</td><td>1745</td></tr>
<tr><td>12:00 midnight</td><td>0000</td></tr>
</table></td><td>On track
• Understands that a day begins at midnight (0000) and midday/noon (1200) is halfway through the day.
• Can confidently convert between 12-hour time and 24-hour time.
• Can correctly use a.m., p.m. and 24-hour notation.
Further learning required
• May confuse 12-hour and 24-hour time notation.
• May be unable to differentiate between a.m. and p.m.
• May not appreciate that 1200 is neither a.m. (before midday) nor p.m. (after midday).</td></tr>
</table>

Topic – benchmarks / Es & Os	Question	Answer	Marking guidance
MNU 2-10a	52	The next four buses are: 0855 – five minutes to nine 0925 – twenty-five minutes past nine 0955 – five minutes to ten 1025 – twenty-five minutes past ten	**On track** • Knows ½ hour = 30 minutes and can add on 30 minutes where it is necessary to bridge an hour. • Can confidently record their solutions in 24-hour time and in words using minutes past/to. **Further learning required** • May confuse 12-hour and 24-hour time notation. • May be unable to add on half an hour when required to bridge an hour, e.g. may add 30 mins to 8:55 a.m. to equal 8:85.
MNU 2-10b	53	A variety of answers are acceptable depending on pupil experience. a Read a book – **hours or days** b Fly to the USA – **hours** c Sneeze – **seconds** d Brush your teeth – **minutes**	**On track** • Can estimate how long an activity would take based on their own experiences and knowledge and understanding of time vocabulary. **Further learning required** • May be confused between seconds, minutes, hours and weeks.
MNU 2-10a	54	Their race lasted **10 minutes 20 seconds**. A variety of strategies are acceptable.	**On track** • Can choose an appropriate method (e.g. empty number line and/or informal jottings) to accurately calculate total duration where it is necessary to bridge minutes. E.g. 4×2 min = 8 min 35 s + 25 s = 1 min 40 s + 20 s = 1 min + 20 s 8 min + 1min + 1min + 20 s = 10 min 20 s **Further learning required** • May use ineffective strategies to calculate duration, e.g. standard algorithm. • May have difficulty bridging minutes when calculating duration, e.g. 40 s + 40 s.

Topic – benchmarks / Es & Os	Question	Answer	Marking guidance
Measurement MNU 2-11a	55	Estimated lengths and difference will vary. Actual lengths are: a 4 cm 8 mm b 12 cm 3 mm c 2 cm 5 mm d 7 cm 1 mm	**On track** • Can make reasonable estimates of the length of the lines and can write these estimations in centimetres and millimetres. They may use their own internal benchmarks to estimate, e.g. I know that my rubber is about 5 cm long. • Can measure the lines accurately with a ruler and write these measurements in centimetres (cm) and millimetres (mm). • Can use their knowledge of place value to calculate the difference between each estimate and actual length. **Further learning required** • May make unreasonable estimates. • May have difficulty using a ruler to accurately measure the length of the lines. This may be because they do not start from the 'zero' point on the ruler or are using an incorrect scale to measure in centimetres and millimetres. • May not be secure in their understanding of place value so may have difficulty calculating the difference, especially when bridging a whole e.g. 4 cm 3 mm – 2 cm 7 mm.
MNU 2-11b	56	Note, some pupils may give their answers in decimal notation a 1400 g and 1 kg 400 g (or 1.4 kg) b 1900 g and 1 kg 900 g (or1.9 kg) c 600 g and 0 kg 600 g (or 0.6 kg) Answers to objects in the environment that weigh approximately the same as the above weights will vary.	**On track** • Can use their knowledge of kilograms and place value to calculate the graduations on the scales. • Can read the scales accurately and record their answers using the correct unit of measurement – grams and kilograms. • Can identify objects in the environment that approximately weigh the same as the measurements given. **Further learning required** • May be unable to calculate what each graduation on the scales shows. • May be unable to read the scales or record the mass of the objects correctly. • May confuse grams and kilograms.

<table>
<tr><th>Topic – benchmarks / Es & Os</th><th>Question</th><th>Answer</th><th>Marking guidance</th></tr>
<tr><td>MNU 2-11b</td><td>57</td><td><table><tr><th></th><th>Capacity in l and ml</th></tr><tr><td>a</td><td>300 ml</td></tr><tr><td>b</td><td>1 litre and 900 ml</td></tr><tr><td>c</td><td>4 litres and 400 ml</td></tr></table></td><td>On track
• Can use their knowledge of millilitres and place value to calculate the graduations on the scales.
• Can read the scales accurately and record their answers using the correct unit of measurement – millilitres and litres.
Further learning required
• May be unable to calculate or understand what each graduation on the scales shows.
• May be unable to read the scales or write down the volume of the liquid correctly.
• May confuse millilitres and litres.</td></tr>
<tr><td>MNU 2-11a</td><td>58</td><td>Answers will vary. Some possibilities are:
a stones – my mass
b feet – my height
c miles – distance from my house to somewhere
d pints – milk</td><td>On track
• Understands the vocabulary used to measure mass, volume and distance and can give examples of when imperial units might be used.</td></tr>
<tr><td>MNU 2-11c</td><td>59</td><td>a The square should have sides 5 cm long. The rectangle has many answers, e.g. 9cm × 1cm, 8cm × 2cm, 7cm × 3cm, 4cm × 6cm</td><td>On track
• Can use their knowledge of the properties of 2D shapes to draw a square and a rectangle with the specified perimeter.
Further learning required
• May not understand the word 'perimeter' or confuse it with area.
• May be unsure of the properties of 2D shapes so may use a trial and error approach to reach 20 cm.</td></tr>
</table>

Topic – benchmarks / Es & Os	Question	Answer	Marking guidance
		b The area of the squares and rectangles will be different, despite having the same perimeter.	**On track** • Can multiply accurately to calculate area and use cm^2 to describe this. **Further learning required** • May not understand the word 'area' or confuse it with perimeter. • May be unable to calculate using the measurements or be unsure about effective strategies to calculate the area, e.g. may not know that $a = l \times b$. • May not use cm^2 as a unit of measurement.
		c The area of one triangle will be 12.5 square centimetres (half the area of the square). The area of the other triangle should be half the area of the rectangle drawn.	**On track** • Can use squared paper to calculate the area of triangles. **Further learning required** • May be confused about the difference between perimeter and area or how to calculate the area of triangles using squared paper.
MNU 2-11c	60	a Nuria and Isla's cuboids both have a volume of $120\,cm^3$. b This is because, even though their base layers look different, they have the same surface area.	**On track** • Can calculate the volume of cuboids by counting cubes/layers of cubes and record using cubic centimetres/cm^3. • Appreciates that objects which look different can have the same volume. **Further learning required** • May confuse volume with perimeter or area. • May believe that objects that look different must have different volumes. • May not use cubic centimetres/cm^3 to describe the volume.

<table>
<tr><th>Topic – benchmarks / Es & Os</th><th>Question</th><th>Answer</th><th>Marking guidance</th></tr>
<tr><td>Patterns and relationships
MTH 2-13a</td><td>61</td><td><table><tr><td>(D)</td><td>1</td><td>2</td><td>3</td><td>4</td><td>5</td></tr><tr><td>(C)</td><td>4</td><td>8</td><td>12</td><td>16</td><td>20</td></tr></table>Formula is:
Number of chairs = 4 × number of desks
C = 4 × D

You would need 32 chairs for 8 desks because 4 × 8 = 32.</td><td>On track
• Can work out what the pattern is and continue it.
• Can write the formula that explains the pattern.
• Can use the formula to work out an unknown when given a known value.
Further learning required
• May be unable to work out or complete the pattern.
• May identify the pattern as 'multiply by 4' / four times table but have difficulty writing a formula to describe it.</td></tr>
<tr><td>Expressions and equations
MTH 2-15a</td><td>62</td><td>a 4 × 6 = 24
b 6 × 5 = 30
c 3 × 4 = 19 – 7</td><td>On track
• Can use known facts and/or knowledge of the inverse relationship to find an unknown number.
• Understands that the equals sign symbolises balance in an equation.
Further learning required
• May be unable to recall basic facts and/ or wrongly assume the unknown is always the product, e.g. may answer '96' (4 × 24) for question a or multiply the answer to 19 – 7 by 4 in c.</td></tr>
<tr><td>2D shapes and 3D objects
MTH 2-16c</td><td>63</td><td><table><tr><td>Circle</td><td>Radius</td><td>Diameter</td></tr><tr><td>A</td><td>5 cm</td><td>10 cm</td></tr><tr><td>B</td><td>30 mm</td><td>60 mm</td></tr></table></td><td>On track
• Understands the vocabulary 'radius' and 'diameter'.
• Can calculate the radius when given the diameter and vice versa.
• Can use a ruler and a compass to draw circles of a given size.
Further learning required
• May connect the words radius and diameter with circles but confuse their meanings.
• May be unable to carry out the appropriate calculations.
• May be unable to use a ruler and compass to draw circles of a given size.</td></tr>
<tr><td>MTH 2-16b</td><td>64</td><td>a cube
b triangular-based pyramid (accept tetrahedron)
c square-based pyramid
d cuboid</td><td>On track
• Can predict and name each 3D object from its net.
Further learning required
• May be unable to match the net to the correct 3D object.</td></tr>
</table>

Topic – benchmarks / Es & Os	Question	Answer	Marking guidance
Angles, symmetry and transformation MTH 2-17a MTH 2-17b	65	Answers will vary but diagram should include 2 reflex angles, 2 acute angles and 2 obtuse angles labelled correctly.	**On track** • Can recall the vocabulary of angles and the properties of acute, obtuse and reflex angles. • Can draw angles of a given type and identify acute, obtuse and reflex angles on their own diagram. • Can estimate the sizes of the angles drawn. • Can measure and record angles accurately using a protractor. **Further learning required** • May confuse acute, obtuse and reflex angles and their properties. • May be unable to complete the diagram with 2 acute, 2 obtuse and 2 reflex angles. • May be unable to identify acute, obtuse and/or reflex angles on their diagram or be unable to use strategies to help them, e.g. rotating the paper to look at the angle from a different perspective. • May be unable to measure and record the sizes of the angles drawn in degrees.
MTH 2-17a	66	a 52 degrees b 11 degrees c 115 degrees d 37 degrees	**On track** • Can use known properties of right angles and straight angles to calculate complementary and supplementary angles. **Further learning required** • May confuse the properties of right and straight angles and so be unable to find complementary or supplementary angles. • May make calculation errors.

Topic – benchmarks / Es & Os	Question	Answer	Marking guidance
MTH 2-17c	67	a **180** degrees lie between north and south b **45** degrees lie between north and north-east c **90 (or 270)** degrees lie between north and west d **135 (or 225)** degrees lie between north and south-west	**On track** • Knows that a full turn equals 360 degrees and can make links with the eight-point compass. • Can calculate the difference between points on an eight-point compass in degrees. **Further learning required** • May be unable to equate each point on an eight-point compass with the corresponding angle from north. • May make mistakes when calculating the difference between points on an eight-point compass.
MTH 2-18a	68	a The vertices of the shape are: (1,8), (2,8), (2,6), (3,6), (3,5), (1,5) b The L shape should be reflected in the horizontal line of symmetry and drawn on the grid. The vertices are: (1,0), (1,3), (3,3), (3,2), (2,2) and (2,0).	**On track** • Can identify the line of symmetry and draw the reflected shape on the squared grid. • Understands the difference between horizontal and vertical symmetry. • Can accurately write the coordinates of the vertices of the original shape and the reflected shape. **Further learning required** • May confuse horizontal and vertical symmetry. • May be unable to reflect the shape around the horizonal axis. • May be unable to write coordinates accurately, e.g. writes (1,0) as (0.1).
MTH 2-19a	69	H, I and X have two lines of symmetry. 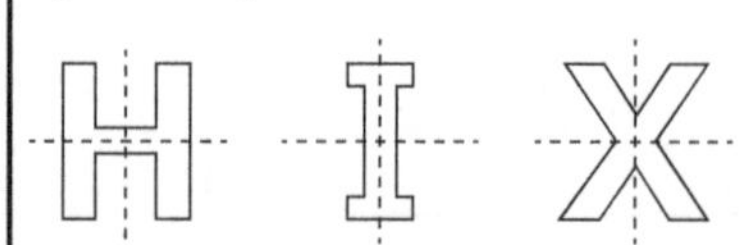	**On track** • Can identify and draw both lines of symmetry on H, I and X. • Understands the difference between horizontal and vertical symmetry. **Further learning required** • May be unable to identify lines of symmetry on the letters given.

Topic – benchmarks / Es & Os	Question	Answer	Marking guidance
MTH 2-19a	70	Check designs for each square tile. Each design should have both vertical and horizontal symmetry.	**On track** • Can identify the horizontal and vertical lines of symmetry. • Can accurately reflect the coloured square pattern in both lines of symmetry. **Further learning required** • May be unable to identify lines of symmetry. • May complete the design inaccurately, e.g. may only use one line of symmetry to complete the design or may simply translate the design.
MTH 2-17d	71	a The scale is 1 cm : 2 km. b 10 km (5 cm on a 1 cm : 2 km scale) c 8 km (4 cm on a 1 cm : 2 km scale) d Answers may vary. May use distances already calculated to estimate this distance.	**On track** • Can confidently use a ruler and a scale to calculate distance on the map. • Can use effective mental maths strategies or informal jottings to calculate distances. **Further learning required** • May be unable to use a ruler correctly to measure the distance between places on the given map. • May have difficulty calculating the actual distance between points on the map using the scale.
Data handling and analysis MNU 2-20a	72	a There were 3000 more of species A than species B in January b April and July showed the least difference between species. There were **1000** of a difference between species in these months.	**On track** • Can confidently read and interpret line graphs. • Can understand word problems and use the data from the line graph to find the solutions. • Can use effective mental maths strategies and informal jottings to make accurate calculations. **Further learning required** • May be unable to interpret the line graph, e.g. has difficulty understanding the axes and/or the scale. • May be unable to interpret the problem.

Topic – benchmarks / Es & Os	Question	Answer	Marking guidance
MNU 2-20a	73	a France was the least popular destination with 50 people choosing to go there. b 150 people went to Spain on holiday. c 40 more people went to Scotland on holiday than did not go away. (Scotland: 100 Did not go away: 60)	**On track** • Knows that a full turn equals 360 degrees and can combine this knowledge and use a protractor to extract data from the pie chart. **Further learning required** • May be unable to work out the number of children represented in each section of the pie chart, either because of a lack of confidence using a protractor or in their methods of calculating.
MNU 2-20a	74	Answers may vary. Some answers may include: a Not reliable – 'Dogs Today' may have bias towards dogs more than other pets to sell more magazines. b Not reliable – Only P6 at Parkmill PS took part in the survey. Other children over Scotland may have responded differently. c Reliable – the NHS is a reliable source who are not a privately-owned company. d Not reliable – Fizzy Drinks company is a manufacturer of fizzy drink so will be promoting fizzy drinks.	**On track** • Can use knowledge of 'big questions' and appropriate statement for inquiry to evaluate the reliability of the data presented. **Further learning required** • May not understand the vocabulary or requirements of the task.
Ideas of chance and uncertainty MNU 2-22a	75	Answers will vary. Statements should be written for the impossible, unlikely, even chance, likely and certain statement that are true for the pupil.	**On track** • Can accurately describe the chance of events occurring in their own lives using the language of probability. **Further learning required** • May be unable to describe events in their life using the vocabulary of chance and probability, e.g. may be able to describe events that are impossible or certain but may have difficulty choosing possible and unlikely events or those that have an even chance.

Yearly progress check 2C

1. a Place 545 674 in its approximate place on the number line below.

b What is 545 674 rounded to the nearest 100 000?

____

c What is 545 674 rounded to the nearest 10 000?

____

2. Are these number statements true or false? Circle true or false. Correct the rounding in the false statements.

a 5·312 rounded to the nearest hundredth is 5·32 **true/false**

b 0·058 rounded to the nearest hundredth is 0·06 **true/false**

c 9·008 rounded to the nearest hundredth is 9·08 **true/false**

3. Isla has worked out this number problem. Check her answer by rounding and estimating and decide if her answer is reasonable. Explain your thinking.

479 928 – 317 304 = 208 624

____

____

4. Write the number shown by these place value houses in both words and numerals.

a

Thousands			Ones		
H	T	O	H	T	O
2	0	6	3	1	0

b

Thousands			Ones		
H	T	O	H	T	O
8	4	5	0	5	8

____

____

5. Finlay made the number 321 152 with place value arrow cards and place value counters.

a What arrow cards would Finlay need to make this number? Write the numerals on the cards to make Finlay's choice correct.

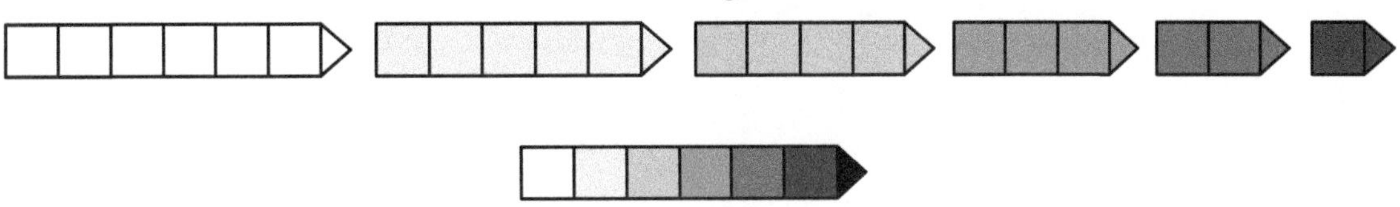

b Finlay lays out 14 place value counters. Which counters did he choose to make 321 152? Write the value on the counters.

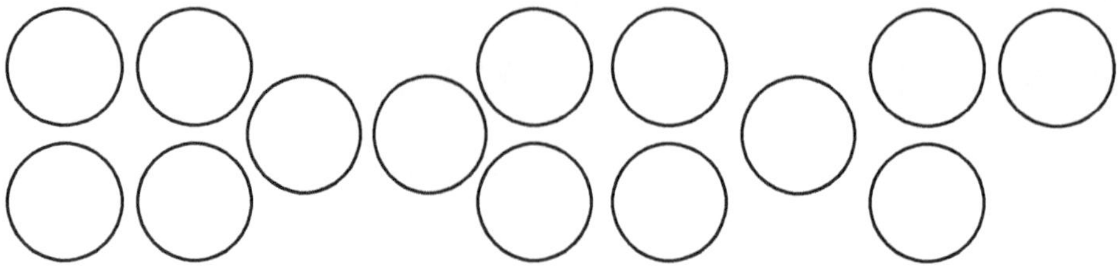

6. Partition the number 210 504 in four different ways.

____

____

____

____

7. Describe the number sequences below. Using words such as increasing or decreasing may help.

a 700 019, 700 009, 699 999, 699 989

____

b 180 145, 190 145, 200 145, 210 145

____

8. Complete each statement by writing < or > in each box.

a 675 504 [ ] six hundred and seventy-five thousand, and fifty-four

b one hundred and one thousand, seven hundred and twenty-three [ ] 110 723

c Using digits, write all four numbers above from smallest to largest.

____

Yearly progress check 2C

9. A diver must complete six dives in a month for her diving qualification. Order the dives she completed starting with the deepest.

−22.5 m   −32.5 m   −10 m   −17 m   −5 m   −22 m

____

10. Write the numbers shown by place value houses **a** and **b** in the following ways:

 i. how you would say the number,

 ii. how many wholes and thousandths the number has,

 iii. the decimal number.

 iv. Describe the position of the place holders in the decimal fractions above.

a

Ones				Decimal Fractions		
H	T	O		$\frac{1}{10}$	$\frac{1}{100}$	$\frac{1}{1000}$
	4	6	•	0	9	4

b

Ones				Decimal Fractions		
H	T	O		$\frac{1}{10}$	$\frac{1}{100}$	$\frac{1}{1000}$
	1	2	•	0	0	9

i. ______________________ ______________________

ii. ______________________ ______________________

iii. ______________________ ______________________

iv. ______________________ ______________________

11. Copy and complete each statement by writing = or ≠ in the box. Explain your answers.

a 5·4 ☐ 5·40   b 0·06 ☐ 0·6   c 0·35 ☐ 0·350

Yearly progress check 2C

12. Write the number represented by these place value counters as:

a a fraction b a decimal number

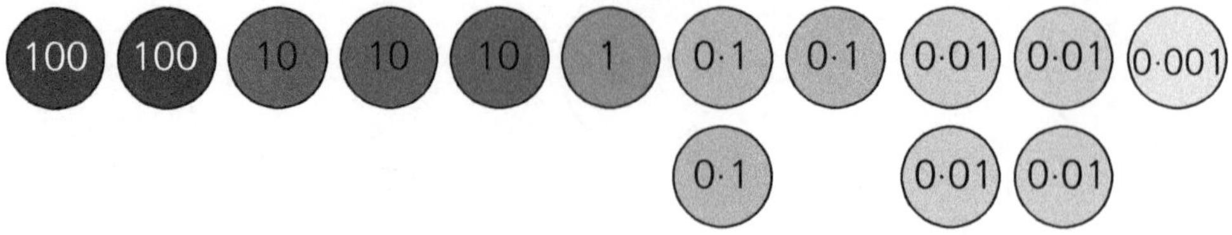

____

____

13. a Write the numbers from largest to smallest.

27·09 27·9 279 7·92 7·927

____

b Finlay says the 8 in 0·083 means 8 hundredths but Amman says it also means 80 thousandths. What do you think?

14. A warehouse holds 948 270 packs of beans. A total of 506 030 packs are put in lorries to go to supermarkets. How many packs of beans are left in the warehouse?

____

15. Calculate the following. Use jottings to help you keep track of your thinking.

a 4700 + 710 + 590 + 1300 + 246 + 34 ____________

b 48 080 – 62 – 18 – 4100 – 900 ____________

16. Calculate the following using a method of your choice:

a 533574 + 365279 ____________ b 194635 − 178127 ____________

Did you use an empty number line, partitioning, an algorithm or a different method? Why was that method chosen?

____

____

17. Isla cannot decide whether to use round and adjust, known facts or a standard algorithm to work out the answers to the following calculations. What advice would you give to Isla? Explain your thinking.

a 150000 + 150000 ____

____

b 30094 − 25000 ____

____

c 180000 − 9999 ____

____

d 3567 + 4672 ____

____

18. Write your own word problem for the calculation 247 139 – 137 139. Use a Think Board to solve your problem.

    Write your answer in a box in the centre of the Think Board.

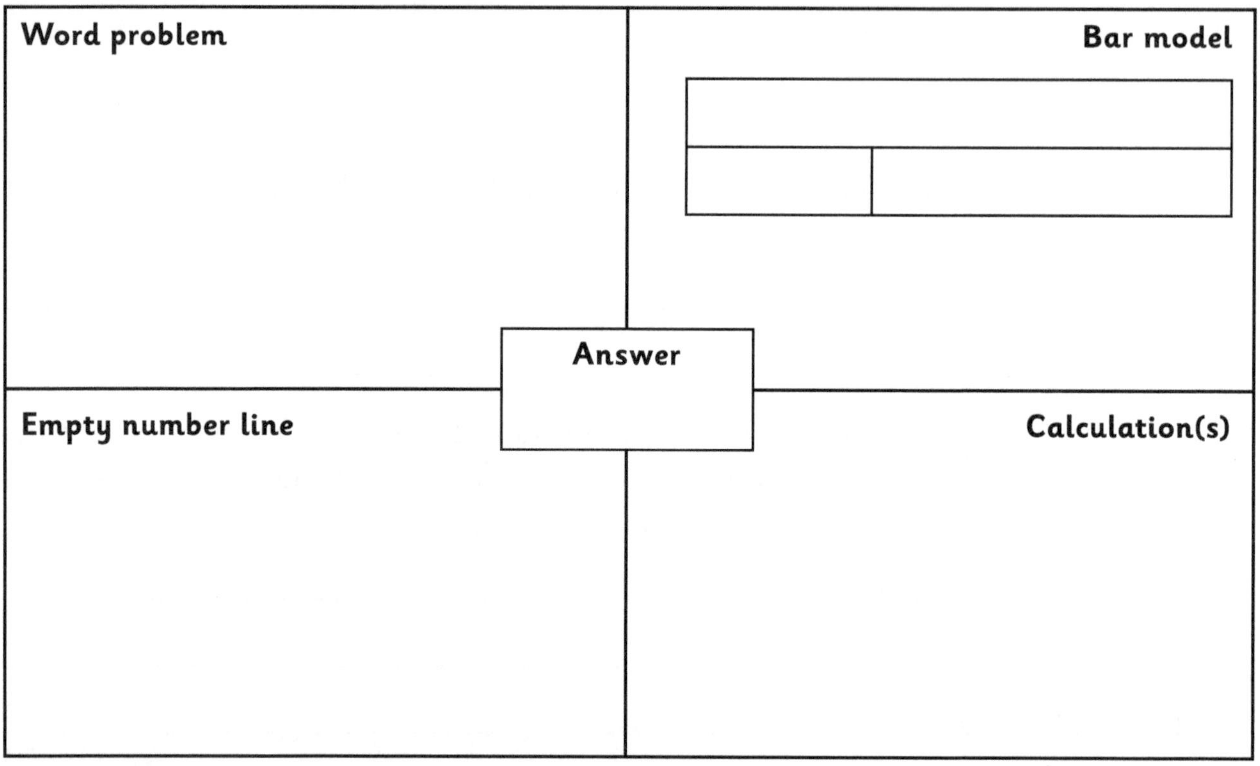

19. a A builder buys a house at auction for £75 600. He spends £34 065 on materials and £15 500 on labour. If he sells the house for £165 000, what was his profit? Show how you worked your answer out.

    b A vet treats 464 patients in one week. $\frac{1}{4}$ of these were cats, 286 were dogs and the rest were small animals. How many small animals did the vet treat? Show how you worked your answer out.

20. a Match pairs of cards to make the decimal fraction 2680·88.

1350·39	1680·43	0·88	2680
80·88	2600	1000·45	1330·49

______________________ ______________________

______________________ ______________________

b Write 3 different decimal fractions that also add up to 2680·88.

____

____

____

21. Use a method of your choice to calculate these addition and subtraction questions.

a 544·52 + 129·88

b 39·32 + 75·67 + 77·81

c 900 – 302·98

d 93·29 – 78·58

e 2000 – 998·41

22. To solve 160·9 – 15·55, would using a mental method or standard algorithm be the most efficient? Explain why and solve the calculation.

23. Amman bought a science kit for £39·99, a magazine for £4·25 and some seeds to plant for 99p. He has £15·29 left. How much money did he have to start with?

    Show how you worked it out.

24. Dad buys three burgers at £4·65 each, four portions of fries at £1·49 each, three milkshakes at £1·60 each and a bottle of water. He gives the cashier two £10 notes and two £5 notes. He gets £3·39 change. How much does the bottle of water cost? Show how you worked it out.

25. Complete these multiplication triangles. Write two multiplication facts and two division facts for each triangle. Show how you worked them out.

a

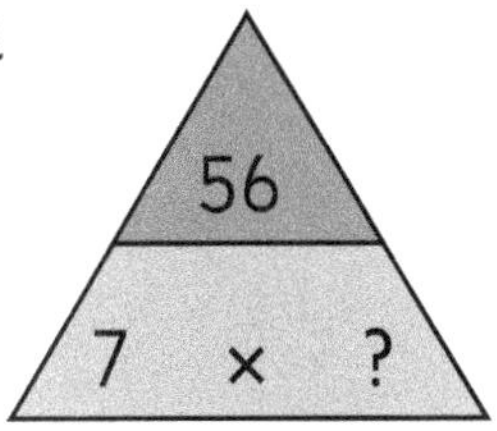

b

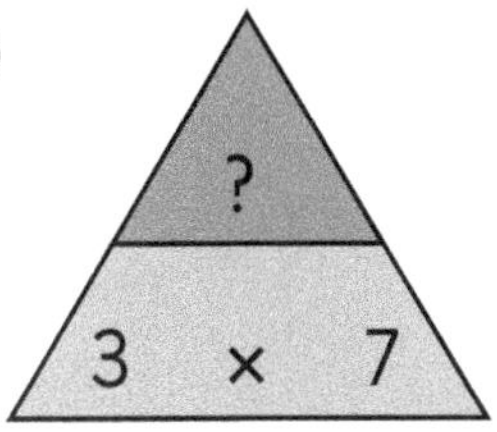

26. Explain how the 2, 4 and 8 times tables are linked.

27. Answer true or false for each of these statements. If false, state what the correct answer should be.

a $7{\cdot}45 \times 100 = 745$

b $1000 \times 3{\cdot}56 = 3000{\cdot}56$

c $0{\cdot}04 \times 10 = 4$

d $267{\cdot}37 \times 100 = 2673{\cdot}7$

28. Isla divides a number by 1000. Next she doubles it, then divides it by 10. Her answer is 0·56. What number did she start with?

29. Use the grid method to answer these multiplication questions:

a 4186 × 6 ______________

b 5651 × 5 ______________

c 3294 × 7 ______________

30. Solve these problems using a strategy of your choice. Explain how you worked each answer out.

a Nuria buys three bottles of juice at £1·79 each.

____

____

b Finlay buys five pairs of socks at £2·35 each.

____

____

c Isla buys six books at £4·67 each.

____

____

31. Calculate the answers to these divisions using a strategy of your choice.

a $552 \div 6$ ______________________

b $702 \div 9$ ______________________

c $344 \div 8$ ______________________

32. Solve these problems using a strategy of your choice. Explain how you worked each answer out.

a Amman earned £18 a week for his paper round.
How much would he earn after seven weeks? ______________________

b Finlay arranged the cakes he had baked into five rows of 12 so that he could ice them.
How many cakes were there? ______________________

33. Use rounding and compensating to solve the following division problem.

a If a factory has made 92 legs, how many tables can be made? Each table needs four legs.

____

b Primary 7 are making small bags of 4 flower bulbs to sell at the school fair. There are 164 bulbs. How many bags can be made?

____

34. Solve these problems using a strategy for your choice. Show how you worked each answer out.

a 14×23 ______

b 16×28 ______

c 7×84 ______

35. Calculate using a method of your choice.

a $135 \div 5$ ______

b $342 \div 8$ ______

c $112 \div 8$ ______

d $2{\cdot}64 \div 4$ ______

e $2{\cdot}88 \div 3$ ______

f $4{\cdot}41 \div 7$ ______

36. a Isla goes shopping and buys two books that cost £9 each and another three that cost £12 each. How much will she have left if she starts with £80? Show how you worked it out.

b Finlay works out the answer to $6 + 8 \div 2$ by adding 6 and 8, then dividing by 2. Will he get the right answer? If not, what has he done wrong and can you correct him?

37. Circle all the numbers that are divisible by 5, 10 or both.

725	150	312	2770	828	1010	1345	816

Explain how you know if a number is divisible by 5 or 10.

38. Use your knowledge of factors and multiples to estimate whose working out is correct. Explain how you know.

Problem	Isla's answer	Finlay's answer	Who is correct and why?
27 × 10	270	271	
135 ÷ 2	65	67·5	
26 × 9	234	270	

39. Use equivalence to compare these fractions. Write the fractions in order from smallest to largest.

$\frac{1}{2}$ $\frac{3}{4}$ $\frac{2}{5}$ $\frac{3}{10}$ $\frac{2}{8}$

____

Yearly progress check 2C

40. Write each fraction in its simplest form.

a $\frac{35}{60}$ b $\frac{60}{100}$ c $\frac{9}{36}$

Draw diagrams on the squared paper to prove your simplified fractions, e.g. that $\frac{25}{100}$ is the same as $\frac{1}{4}$.

41. Solve the following problem. You may wish to draw a diagram or a bar model to help you.

Isla cycles $2\frac{1}{2}$ miles and then runs $1\frac{3}{4}$ miles. How far did she travel?

42. Complete each bar model to show the decimal fraction. Then write each decimal number as a fraction in its simplest form and as a percentage.

0.75 = [ ] = ______________ = ______________
?

0.4 = [ ] = ______________ = ______________
?

0.15 = [ ] = ______________ = ______________
?

43. Solve these problems. It may help to draw a bar model.

a Nuria has half a bag of sweets. She gives a third of her sweets to Finlay. What fraction of the whole bag of sweets does Finlay have?

____

b Isla has one third of a bottle of water. She gives half of her water to Amman. What fraction of the whole bottle does she have left?

____

44. Amman needs one eighth of a cup of flour to make a scone. He has 3 ½ cups of flour. How many scones can he make? Show how you worked it out.

45. Isla has five bars of chocolate. How many portions does Isla get if she divides each bar into eighths? Show how you worked it out.

46. 15% of people who viewed a video online 'liked' the video. How many 'likes' has the video received if it has had 6500 views? Explain how you worked this out.

47. Nuria's mum takes the bus to and from work. The bus company have different options for ticket prices:

A single ticket costs £1·35

A return ticket costs £2·25

A book of four return tickets costs £8·50

Nuria's mum works five days a week. Which tickets should she buy to make her travel as cheap as possible? Show how you worked it out.

48. Finlay's gran earns £1150 per month. Her mortgage is £450 per month and she spends £40 per week on food, assuming there are 4 weeks in a month. How much does she have left for other essentials?

49. Amman helped his dad recover an old sofa.

- His dad bought the sofa for £50 in an auction.
- He needed 3 m of fabric to recover it, at a cost of £8·50 per metre.
- He also had to buy some materials that cost him £12.

If they resold the sofa for £120, how much profit did they make?

50. Isla wants to buy some flowers for her grandpa's birthday. As a new customer, she has the choice of either £5 off, or a 15% discount. If her grandpa's flowers cost £30, which is the better offer? Explain your choice.

51. Nuria's family need to replace their washing machine. They find one for £240 which they can buy using a hire purchase agreement.

If they don't pay a deposit, their monthly payments are £35 for a year. How much more will they pay for the washing machine if they choose hire purchase instead of paying the full amount in cash? Explain your thinking.

52. Finlay's bus leaves the station at 1325 and arrives at its destination 3 hours and 50 minutes later. What time did Finlay arrive?

Yearly progress check 2C

53. A competition is open from 9 a.m. on Monday until 5 p.m. on Friday. For how many days and hours is the competition open?

____

54. Use the formulae to answer the questions below:

Time = Distance ÷ Speed

Distance = Speed × Time

a Find the time taken by a cyclist travelling at an average speed of 20 kilometres per hour (km/h) over a distance over 80 km.

____

b Find the distance travelled by a car driving at approximately 50 miles per hour (mph) for 3 hours.

____

55. This table records the readings from a stopwatch. Complete the table.

Start time	Finish time	Time elapsed (duration)
3 minutes 24 seconds	4 minutes 15 seconds	
5 minutes 46 seconds	10 minutes 24 seconds	
12 minutes 15 seconds		3 minutes 50 seconds

56. a Change 458 cm into m ________________

b Change 1·4 km into m ________________

c Change 3·4 cm into mm ________________

d Change 5310 m into km ________________

Yearly progress check 2C

57. Amman walks to the local shop to buy some vegetables. The scales below show the weight of the carrots, potatoes and broccoli he bought.

Carrots   Potatoes   Broccoli

What is the total weight of Amman's shopping?

58. Find the missing lengths for each of these surface areas.

Mobile phone screen:

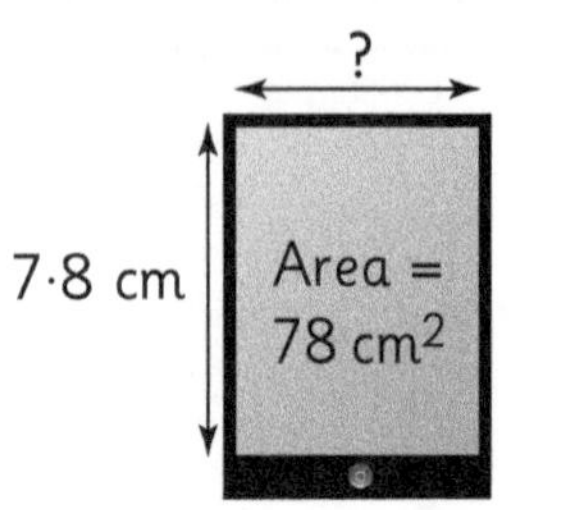

Tablet screen:

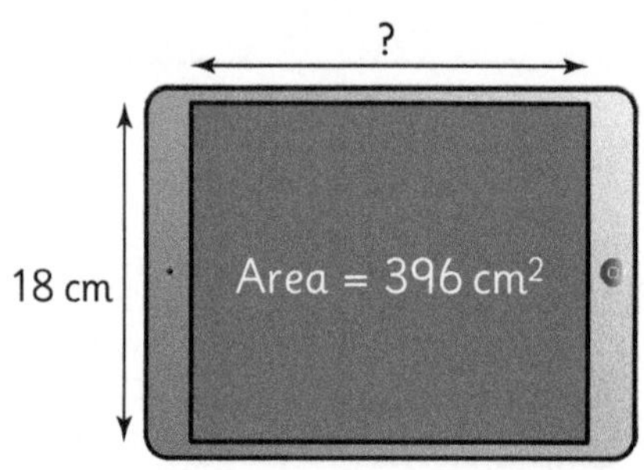

Garden:

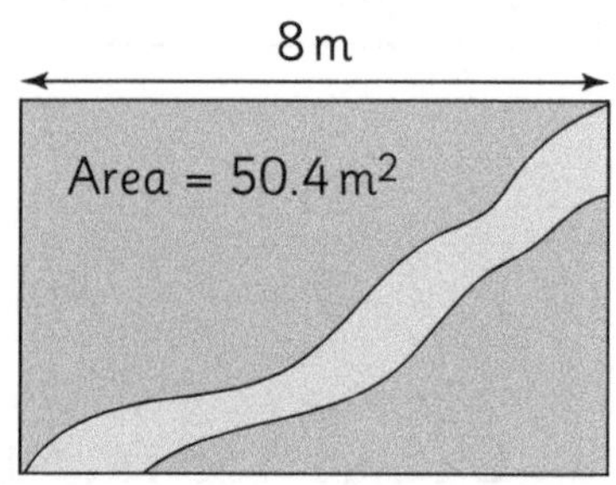

59. Write each of the measurements in both millilitres and litres, e.g. 100 ml and 0·1 l.

a

cranberry juice

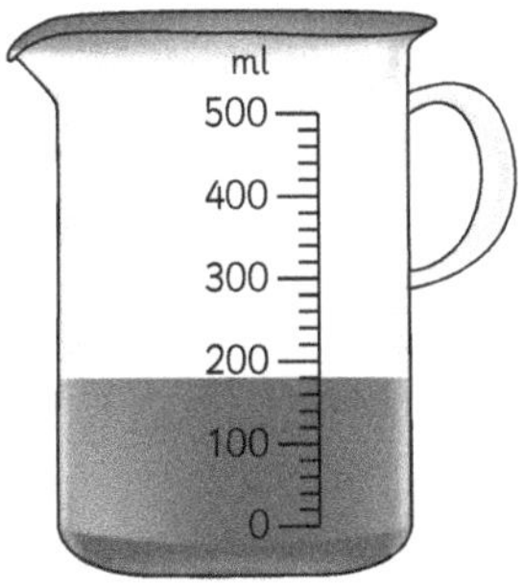

b

apple juice

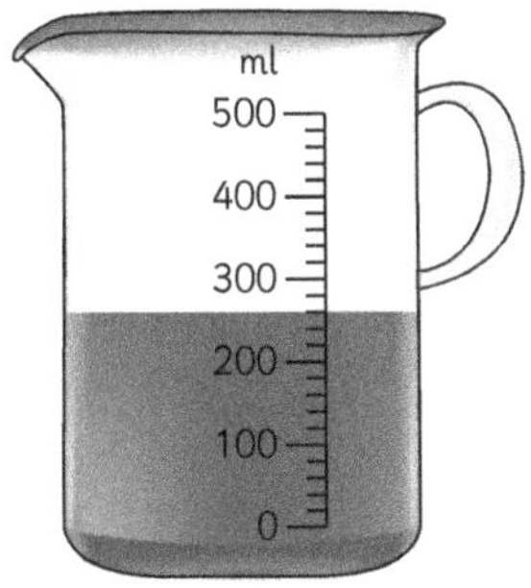

c

orange juice

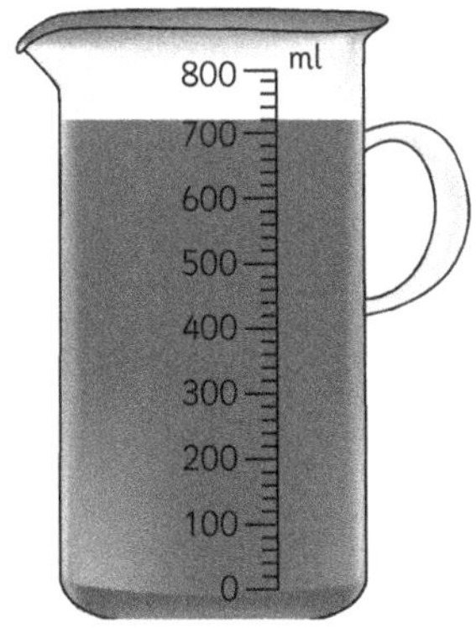

________________ ________________ ________________

60. Complete the table to show your understanding of which units of measurement might be used if using metric and imperial measurements.

Item being measured	Metric units that could be used	Imperial units that might be used
The distance between Edinburgh and Aberdeen		
The height of an adult		
The mass of an adult		
The capacity of a container of milk		

Yearly progress check 2C

61. a Sketch a plan of your ideal playground. Include at least three different zones that have straight sides and that have perimeters of 50 m, 32 m and 18 m. Ensure that the lengths of the sides are included in your sketch. Your shapes must be squares, rectangles, L shapes or crosses.

b Work out the area of each of the playground zones. Use jottings to help you keep track of your thinking.

62. Calculate the area of these right-angled triangles. Be sure to show your working.

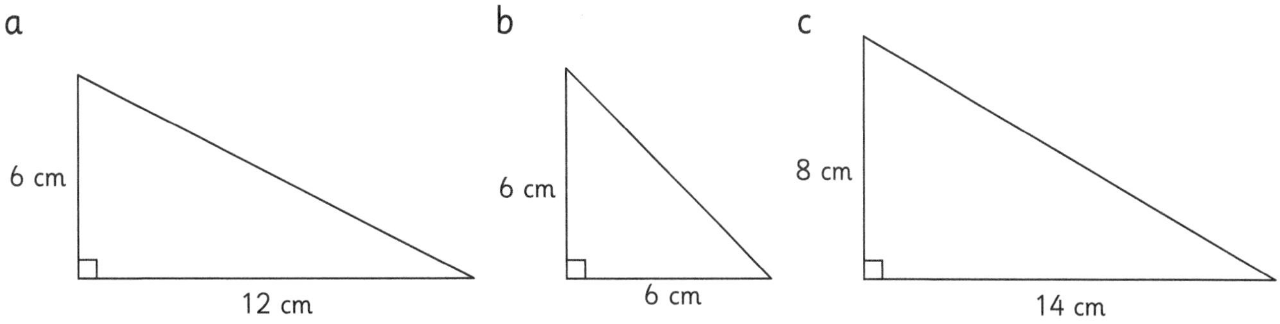

63. This table shows the Babylonian number system. Write the factors of 19 in Babylonian.

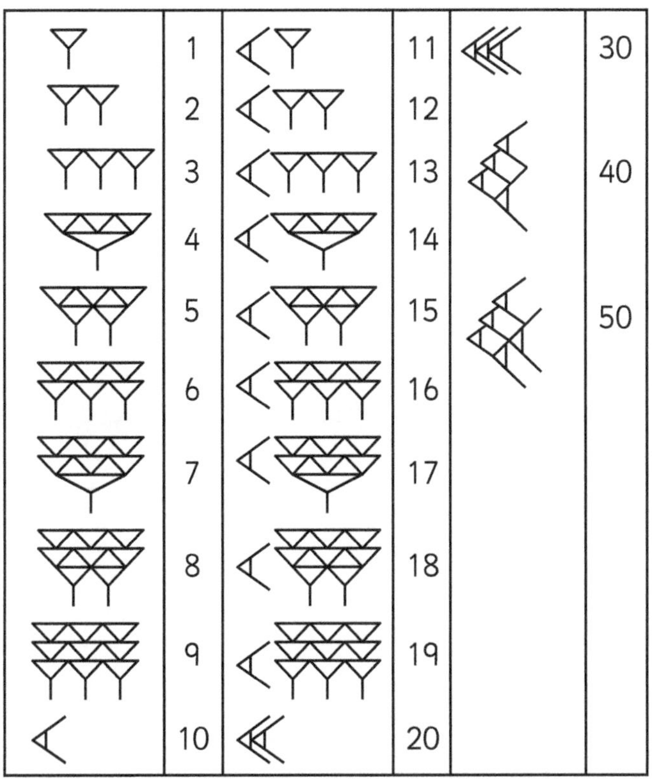

64. Continue the following number sequences and identify whether these are square or triangular numbers. Draw diagrams to prove your answers.

a 1, 3, 6, 10, ______, ______

b 1, 4, 9, 16, ______, ______

Yearly progress check 2C

65. Find the missing numbers.

a Find the value of ▲ if ● = 6

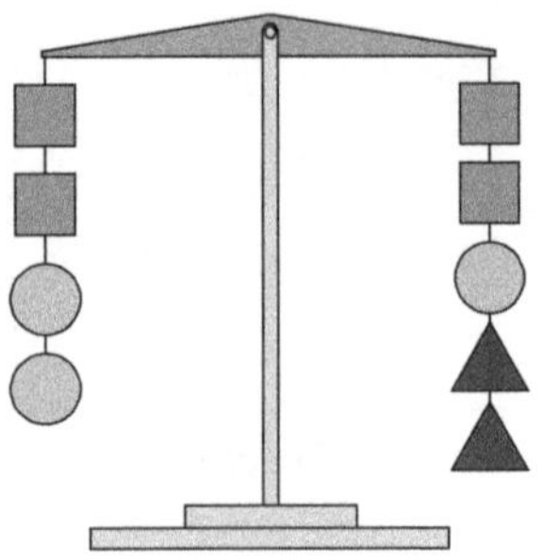

b Find the value of ● if ▲ = 8

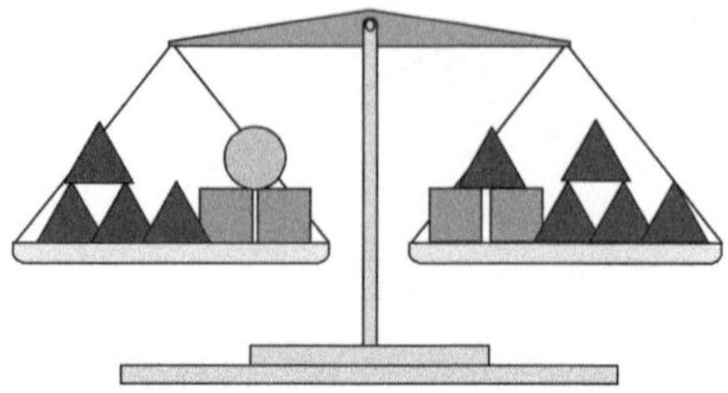

66. Use a ruler and protractor to draw the following triangles:

a a scalene triangle with one side 8 cm and one angle of 60°

b an isosceles triangle with a base that is 6 cm long

c an equilateral triangle

Yearly progress check 2C

67. Sketch two different examples of each of these shapes: rectangle, square, equilateral triangle, parallelogram, isosceles triangle, trapezium.

The shapes should be in different positions. Write the name of each shape beside your sketches.

Three-sided shapes	Four-sided shapes	

68.

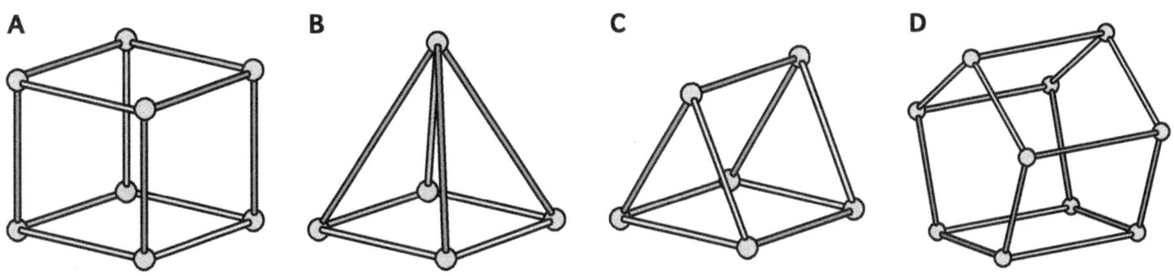

These are skeletal models of 3D objects. Which models have:

a more than three triangular faces? ______

b more than three edges at one vertex? ______

c three edges at one vertex? ______

d more than four rectangular faces? ______

Yearly progress check 2C

69. Name the 3D objects from their nets.

a

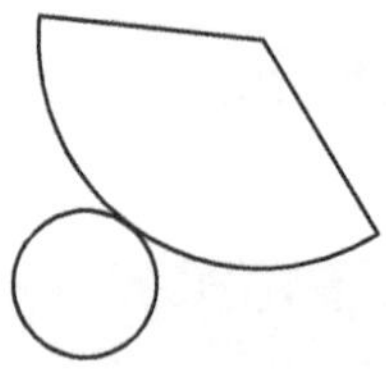

b

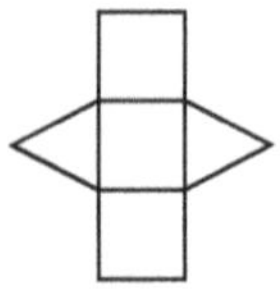

c

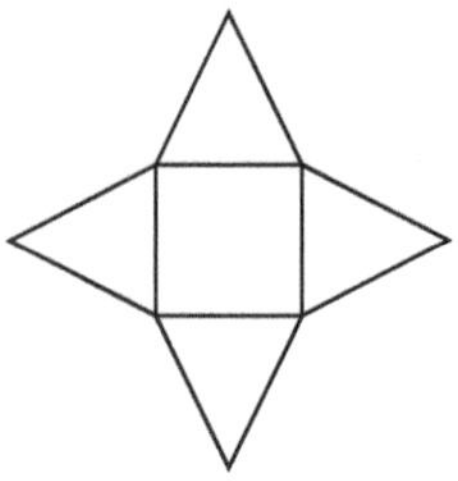

d

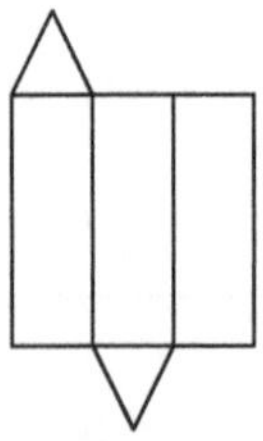

70. Measure the missing lengths and angles from these triangles.

a

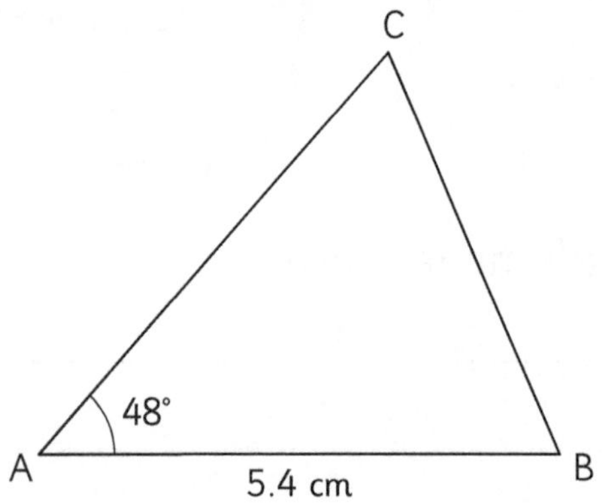

b

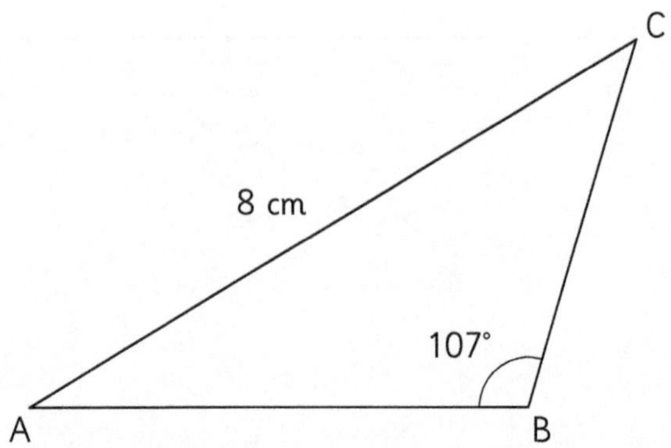

Yearly progress check 2C

71. For each of these reflex angles, calculate the acute or obtuse angle.

a

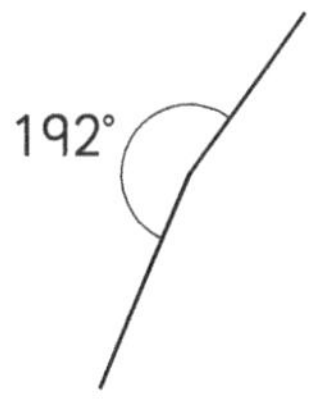

____

b

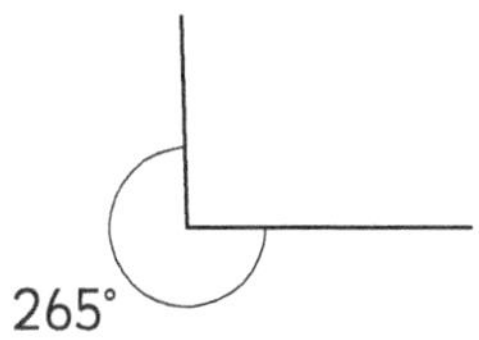

____

c

____

72. Without measuring, calculate the missing angles.

a

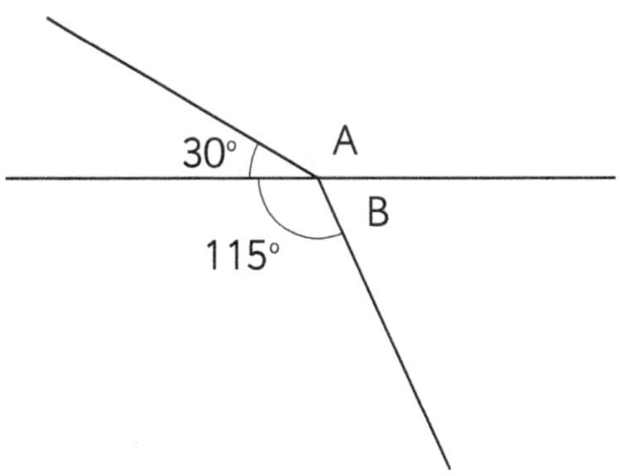

A ____________ B ____________

b

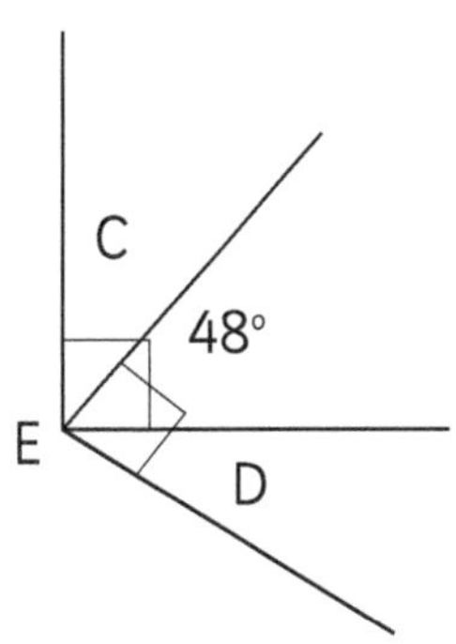

C ____________ D ____________ E ____________

73. Nuria is standing facing north clockwise.

She turns 315° clockwise.

What compass direction is she now facing? ______________________________

74. The pair of shapes below is congruent. Find the missing coordinates.

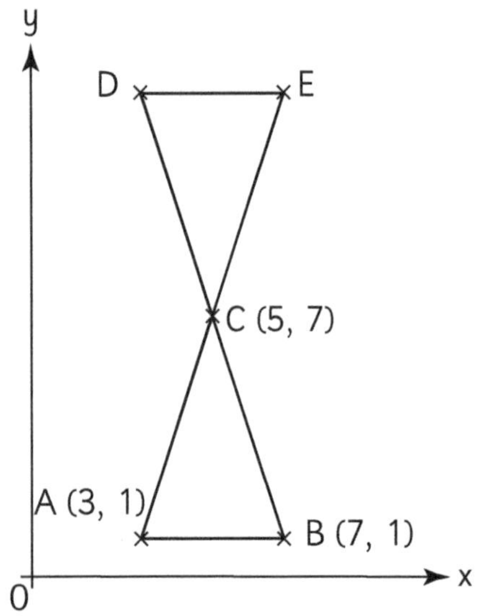

75. a Sort these shapes according to their lines of symmetry and number of sides. Put the letter of each shape in the correct place in the Carroll diagram.

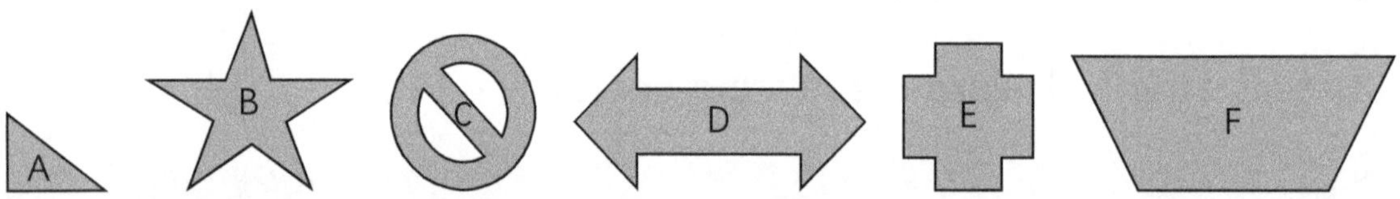

	One or more diagonal line of symmetry	**Vertical or horizontal lines of symmetry**
6 sides or more		
5 sides or fewer		

b Which shape doesn't belong in the table? Explain your thinking.

76. Create a design so that it has 4 lines of symmetry.

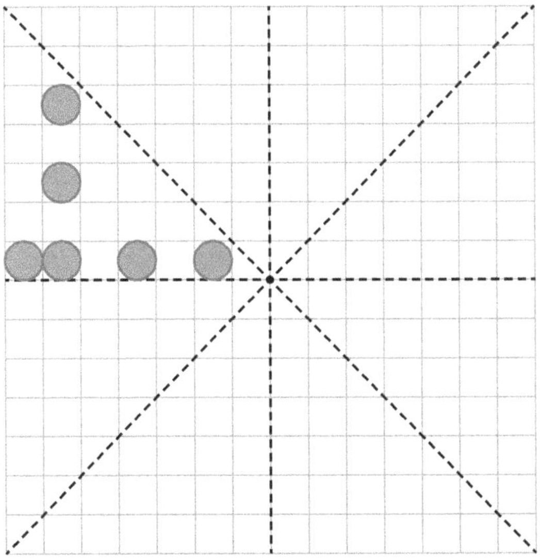

77. Finlay has been travelling around Scotland. He has written the distances between some of the places he has been to.

	Places in Scotland	Distance in km
a	Edinburgh Castle, Edinburgh – The Kelpies, Falkirk	45
b	The Kelpies, Falkirk – Wallace Monument, Stirling	30
c	Culloden Battlefield – Dunrobin Castle, Golspie	90
d	John O'Groats – Ullapool	190

Finlay decides on a scale of 1 cm : 10 km.

Calculate then draw a straight line to show each distance as it would appear on a map, e.g.

Don't forget to label the places!

a

b

c

d

Yearly progress check 2C

78. a Write two explanations of why the survey below DOES NOT gather data from all the people involved (population).

On Friday, all Primary 7 pupils at Anytown Primary were asked their views on the school's after-school clubs.

b Suggest a more appropriate sample for the school to use to gather information about after-school clubs.

79. Three companies are communicating their data using graphs. What conclusions can you draw from the graphs? Why do you think the companies have decided to present the data like this?

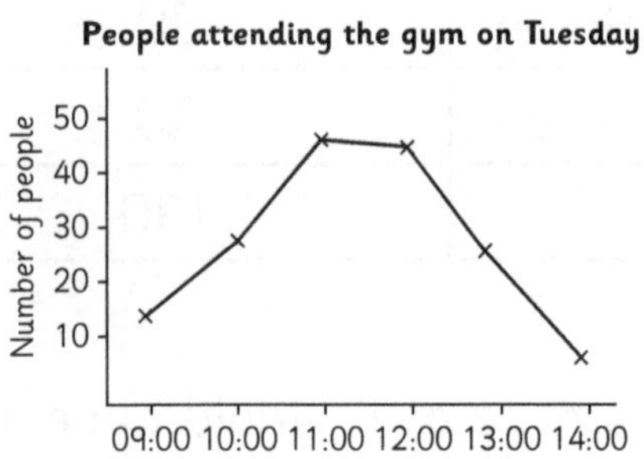

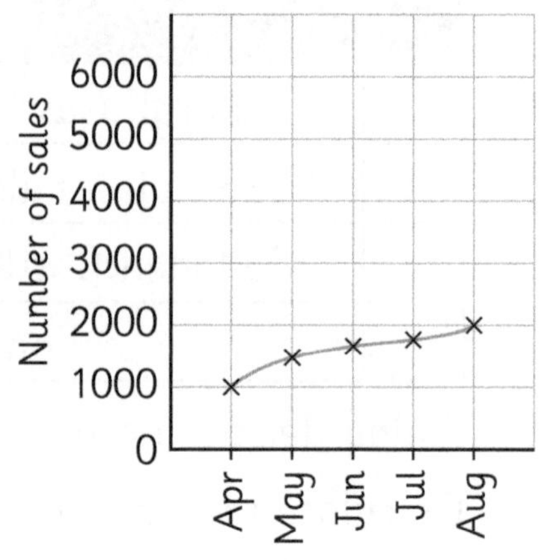

Yearly progress check 2C

80. Amman and Isla record the sales for three products in their healthy tuckshop for one month.

   a Complete the table.

Product	Number sold	Number of degrees	Percentage of the total
Porridge	180		
Melon	108		
Apple	72		
TOTAL:	360	360	100%

   b Complete the pie chart. Label it with the percentages.

Pie chart

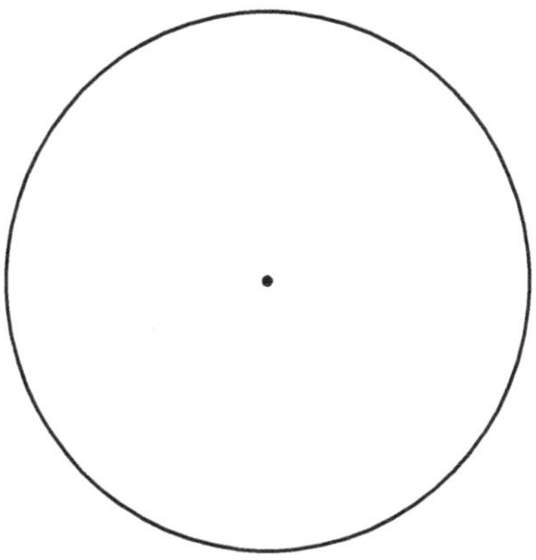

   c What do you notice about the data in the pie chart?

____

____

81. You have two dice.

   a If you roll two standard dice and add the scores together, what is the likelihood that you will get an odd number?

____

   b A standard dice is rolled 72 times. How many times would you expect to roll a 3?

____

Topic – benchmarks / Es & Os	Question	Answer	Marking guidance
Estimation and rounding MNU 2-01a	1	a 545 674 should be placed to the left of the mid-point (close to 550 000) on the number line. b 545 674 is 500 000 to the nearest 100 000 c 545 674 is 550 000 to the nearest 10 000.	**On track** • Understands the conventions of rounding – rounding up when the digit you are rounding at (known as the rounding digit) is 5 or more, rounding down when the digit is 4 or less. • Can round to the nearest 10 000 and 100 000 accurately. **Further learning required** • May use incorrect rounding strategies, e.g. focusing on the incorrect digit for rounding or rounding to the nearest hundred or ten, instead of to the nearest 100 000 or 10 000.
MNU 2-01a MNU 2-02a	2	a False: 5.312 rounded to the nearest hundredth is 5.31 b True c False: 9.008 rounded to the nearest hundredth is 9.01	**On track** • Understands the conventions of rounding, rounding up when the rounding digit is 5 or more, rounding down when the digit is 4 or less. • Can round decimal fractions accurately to the nearest hundredth. **Further learning required** • May confuse when to round up or round down, e.g. may base their decision to round up or down on the hundredths, rather than the thousandths, digit.
MNU 2-01a	3	If rounding to the nearest 1000, the calculation becomes 480 000 – 317 000 = 163 000. If rounding to the nearest 1000 or 10 000, the calculation becomes 480 000 – 320 000 = 160 000. This means that Isla's calculation is incorrect.	**On track** • The pupil can reason about the numbers and identify how best to round the values. • Understands the conventions of rounding – rounding up when the rounding digit is 5 or more, rounding down when the digit is 4 or less. • Can round to the nearest 1000 and 10 000 accurately. • Can subtract using effective strategies. **Further learning required** • May use incorrect rounding strategies, e.g. focusing on the incorrect digit for rounding or rounding to the nearest hundred instead of the nearest thousand. • May not understand how rounding can be used to estimate the answer to calculations. • May choose to subtract the two numbers instead of rounding.

Topic – benchmarks / Es & Os	Question	Answer	Marking guidance
Number – order and place value MNU 2-02a	4	a 206 310 – two hundred and six thousand, three hundred and ten b 845 058 – eight hundred and forty-five thousand, and fifty-eight	**On track** • Understands the purpose of zero as a place holder. • Writes number words accurately. Although accurate spelling should be encouraged, reasonable and recognisable attempts (e.g. hundrid) should be accepted. **Further learning required** • May not understand zero as a place holder and/or attach a value to it, e.g. may (wrongly) believe that all zeros can be omitted and writes 2631 instead of 206 310. • May write six-digit number words incorrectly, e.g. may omit the 'thousand' or 'and' in 'two hundred and six thousand'.
MNU 2-02a	5	a Finlay uses the following arrow cards: 300 000, 20 000, 1 000, 50 and 2 b Finlay uses the following 14 place value counters: 3 × 100 000 2 × 10 000 1 x 1000 1 x 100 5 × 10 2 × 1	**On track** • Understands that each card or counter has a value attached to it. • Understands the relationship between the arrow cards, place value counters and the number they represent. • Understands standard place value partitioning. **Further learning required** • May be unfamiliar with arrow cards or place value counters. • May use place value arrow cards and a 'trial and error' approach in part a and/or write the incorrect digits on the counters.
MNU 2-02a	6	Answers may vary.	**On track** • Flexibly partitions 210 504 in four different ways, which shows understanding of standard partitioning (200 000 + 10 000 + 500 + 4) and non-standard partitioning, e.g. 210 000 + 500 + 4; 100 000 + 110 000 + 504 etc. **Further learning required** • May be unable to partition 210 504 according to the place value of the digits or have difficulty 'seeing' 210 504 as anything other than 200 000 + 10 000 + 500 + 4.

Topic – benchmarks / Es & Os	Question	Answer	Marking guidance
MNU 2-02a	7	a Decreasing by 10 b Increasing by 10 000	**On track** • Identifies the pattern/rule used to generate each sequence and describes it accurately. **Further learning required** • May be able to say that the pattern is going down/decreasing, or going up/increasing, but be unable to identify by how much. • May have difficulty with numbers that bridge 100 000. • May find backwards number sequences more challenging that forward number sequences.
MNU 2-02a	8	a 675 504 **>** six hundred and seventy-five thousand, and fifty-four b one hundred and one thousand, seven hundred and twenty-three **<** 110 723 c 101 723, 110 723, 675 054, 675 504	**On track** • < and > and are used to complete the expressions correctly. • Can write six-figure number words as digits. • Can order six-figure numbers from smallest to largest. **Further learning required** • May mix up the inequality symbols < and >. • May be unable to write the numbers displayed in words correctly in numerals. • May be unable to order the numbers from smallest to largest.
MNU 2-04a	9	−32.5 m, −22.5 m, −22 m, −17 m, −10 m, −5 m.	**On track** • Orders depth correctly from deepest to shallowest. **Further learning required** • May not understand the meaning of 'ascending order'. • May understand the phrase 'ascending order' but wrongly believe that −5 m should come first because 5 is the smallest number.

Topic – benchmarks / Es & Os	Question	Answer	Marking guidance
MNU 2-02a MNU 2-03b	10	a i) forty-six point zero nine four ii) 46 wholes and ninety-four thousandths iii) 46.094 iv) tenths b i) twelve point zero zero nine ii) 12 wholes and 9 thousandths iii) 12.009 iv) tenths and hundredths	**On track** • Correctly identifies the decimal fraction (wholes and thousandths) represented by each place value house. • Writes each decimal fraction correctly in words and numerals. Although accurate spelling should be encouraged, reasonable and recognisable attempts should be accepted. **Further learning required** • May be confused about converting the numbers in the place value houses to decimal fractions, either because they are unfamiliar with the fractions used to represent hundredths and thousandths or due to a lack of understanding about using zero as a place holder. • May not understand the place value relationship between hundredths and thousandths. • May be unable to write decimal fractions accurately in words, e.g. 46.094 may be written as forty-six point ninety-four.
MNU 2-02a	11	a $5.4 = 5.40$ b $0.06 \neq 0.6$ c $0.35 = 0.350$	**On track** • $=$ and $\neq$ are used to complete each expression correctly. **Further learning required** • May be confused about the difference between $=$ and $\neq$. • May believe that 5.4 is not the same as 5.40, e.g. may be able to read 0.350 as 350 thousandths but not equate this with 35 hundredths. • May believe that the more digits there are the greater the number and vice-versa.
MNU 2-02a	12	a $231 \frac{341}{1000}$ b 231.341	**On track** • Understand that the place value counters represent the number two hundred and thirty-one point three four one. • Correctly writes this number as a fraction and as a decimal. **Further learning required** • May be unable to relate the place value counters to a decimal fraction.

Topic – benchmarks / Es & Os	Question	Answer	Marking guidance
MNU 2-02a MNU 2-03b	13	a 279, 27.9, 27.09, 7.927, 7.92 b Answers may vary but both Finlay and Amman are correct – the digit is 8 hundredths or 80 thousandths.	**On track** • Orders the decimal fractions correctly from largest to smallest. • Can explain in words, or show by drawing, why an '8' in the hundredths place can have a value of both 8 hundredths and 80 thousandths. **Further learning required** • May disregard the zero after the decimal point so believe that 27.9 and 27.09 are equal. • May order the numbers according to the value of the digits immediately after the decimal point and so believe that 27.9 is the largest number because it only has 'tenths'. • May believe that 7.927 is the largest number because it has the most digits after the decimal point.
Number – addition and subtraction MNU 2-02a MNU 2-03a	14	442 240 packs	**On track** • Identifies subtraction as the correct operation and calculates accurately using a method of their choice. **Further learning required** • May be unable to identify the operation required to solve the problem or chooses the wrong operation, e.g. addition instead of subtraction. • May calculate inaccurately; possibly due to choosing an appropriate strategy but using it incorrectly.

Topic – benchmarks / Es & Os	Question	Answer	Marking guidance
MNU 2-02a MNU 2-03a	15	a 7580 b 43 000	**On track** • Calculates accurately. Attention should be paid to how numbers are grouped to potentially make calculating easier, e.g. finding numbers that total a multiple of 10/100/1000. • Understands that the numbers can be added in any order without affecting the total and chooses the most efficient method, e.g. for a 4700 + 1300 = 6000, 710 + 590 = 1300 and 246 + 34 = 280 so the solution is 6000 + 1300 + 280 = 7580. • For b understands that the numbers to be subtracted can be combined to ease the calculation load, e.g. 48 080 – 62 – 18 – 4100 – 900 is 48 080 – 5080. • Can use a range of mental strategies (e.g. compensation, make 10/100/1000), supported by informal jottings, and may choose different strategies, depending on the numbers involved. **Further learning required** • May calculate inaccurately; possibly due to incorrect recording of the numbers in the calculation, the strategy chosen or choosing an appropriate strategy but using it incorrectly. • May choose an inefficient strategy, e.g. jumping back in 10s when subtracting 131. • May complete the calculation in the exact order in which the numbers are written.
MNU 2-03a	16	a 898 853 b 16 508 An explanation should be provided as to why a particular method was chosen.	**On track** • Calculates accurately. Attention should be paid to the appropriateness and efficiency of the strategy or strategies chosen. • May choose a different strategy/method for each calculation. • Can justify their choice of strategy/method. **Further learning required** • May calculate inaccurately; possibly due to incorrect recording of the numbers in the calculation. • May choose an appropriate strategy/method but use it incorrectly. • May choose an inefficient strategy, e.g. tries to jump back in 10s, 100s or 1000s when adding or subtracting 6-digit numbers. • May be over reliant on the same, one strategy.

Topic – benchmarks / Es & Os	Question	Answer	Marking guidance
MNU 2-02a MNU 2-03a	17	a Known facts – we know that 15 + 15 = 30 so we can use that to work out 150 000 + 150 000. b Round and adjust – Subtract 25 000 from 30 000 then add on 94. c Round and adjust – Subtract 10 000 then add on 1. d Algorithm – this calculation involves quite complex numbers so a standard algorithm may be easiest. Alternatively, the pupil may choose to use place value partitioning and a semi-formal written method, recording the partitions horizontally or vertically.	**On track** • Can use and apply knowledge of how numbers are constructed (place value) to identify the most efficient strategy/method for each calculation. • Can explain the reasoning behind their choice of strategy/method. **Further learning required** • May rely on vertical algorithms to solve each calculation. • May be unable to determine the most appropriate method. • May be over reliant on the same, one strategy that may not necessarily be the most efficient in every case, e.g. standard written algorithm.
MNU 2-03a	18	The word problem created should link clearly with the result unknown subtraction given, i.e. 247 139 – 137 139 = ? A Think Board is constructed and includes the sections: 'Word Problem', 'Bar Model', 'Empty Number Line' and 'Calculations'. The solution (110 000) is placed in the centre.	**On track** • Can write a word problem to fit with the calculation given. • Can confidently complete all parts of the Think Board. These should relate clearly to the word problem the pupil has written. • Can flexibly partition numbers and use an effective strategy (e.g. place value partitioning) to calculate the answer. **Further learning required** • May be unable to write a word problem to match the given calculation. May include the numbers 247 139 and 137 139 but not be able to relate the 'story' to a result unknown subtraction. • May only be able to complete some parts of the Think Board. • May use an ineffective strategy to calculate the solution.

Topic – benchmarks / Es & Os	Question	Answer	Marking guidance
MNU 2-03a MNU 2-09c	19	a The builder makes a profit of £39 835 (total cost of house, materials and labour is £125 165 and he sold the house for £165 000). b The vet treats 62 small animals (cats = 116, dogs = 286 and the total of animals treated is 464)	**On track** • Can solve both word problems, clearly showing and/or describing how they calculated each step. • Can choose the correct operation (addition, subtraction, division) for each step. • Can choose an effective strategy for each calculation, based on the numbers involved. • Can flexibly partition the numbers in order to calculate. **Further learning required** • May be unable to visualise or represent the word problems. • May be unable to connect the various steps of the problems with the appropriate operations/calculations or chooses the wrong operations/calculations. • May not check their solution or use estimation to see if the answer is reasonable. • May default to the same strategy (e.g. standard written algorithms) for every calculation, even where an alternative, mental strategy may be more efficient. • May complete part of the word problem correctly but be unable to find the final solution.
MNU 2-03b	20	a 1000.45 + 1680.43 0.88 + 2680 2600 + 80.88 1350.39 + 1330.49 b Answers will vary.	**On track** • Can use knowledge of standard and non-standard place value partitioning to find pairs of numbers that total 2680.88. **Further learning required** • May only be able to identify the most obvious pairs, e.g. 0.88 + 2680 or 2600 + 80.88. • May be unable to write three decimal fractions that total 2680.88.

Topic – benchmarks / Es & Os	Question	Answer	Marking guidance
MNU 2-02a MNU 2-03b	21	a 674.4 b 192.8 c 597.02 d 14.71 e 1001.59	**On track** • Can calculate accurately. • Is confident using a variety of strategies and methods, based on the numbers involved. • Can flexibly partition the numbers in order to calculate. **Further learning required** • May make simple calculation errors. • May not understand that they can count on or back to subtract. • May default to the same one strategy for each calculation, e.g. vertical algorithms. • May use strategies such as compensation/ round and adjust incorrectly, e.g. may subtract 303 for c but then not add 0.02.
MNU 2-02a MNU 2-03b	22	The answer is 145.35.	**On track** • Can calculate accurately and explain their mental method. • Can use knowledge of place value to flexibly partition the numbers in order to calculate, e.g. 160 – 15 = 145; 90 hundredths – 55 hundredths = 35 hundredths. **Further learning required** • May be unable to calculate mentally and choose to use a standard algorithm. • May make simple calculation errors when subtracting multiples of 5 and 10, e.g. 160 – 15 = 155; 90 – 35 = 45. • May not equate 9 tenths with 90 hundredths and so think it is impossible to subtract point five five from point nine, or may give an answer of 145.46.

Topic – benchmarks / Es & Os	Question	Answer	Marking guidance
MNU 2-03b	23	Amman started off with £60.52.	**On track** • Can interpret the problem and identify the appropriate steps required to solve it, i.e. the pupil knows to add the total cost to the change given. • Can choose and use an effective strategy or method to calculate each step. For example, the child may use round and adjust to find the total spent (£40 + £1 + £4.25 – 2p) then use a jump or split strategy on an empty number line to find how much Amman had to start with. **Further learning required** • May be unable to interpret, visualise or represent the word problem. • May include the numbers in the word problem but choose the wrong operation(s). • May use ineffective strategies to complete the calculations involved.
MNU 2-03b	24	The bottle of water costs £1.90. Methods of working this out will vary but: 3 burgers @ £4.65 each = £13.95 4 fries @ £1.49 each = £5.96 3 milkshakes @ £1.60 each = £4.80 Dad has a total of £30.	**On track** • Can interpret the multi-step word problem and carry out the necessary steps to solve it. • Can choose and use an effective strategy/method to calculate each step. **Further learning required** • May struggle to visualise the word problem and make links with the appropriate operations/calculations. • May use ineffective strategies to complete calculations, e.g. may default to the same strategy for every calculation regardless of the numbers involved. • May answer part of the word problem but not all of it.

Topic – benchmarks / Es & Os	Question	Answer	Marking guidance
Number – multiplication and division MTH 2-03a	25	a 8 $7 \times 8 = 56$ $8 \times 7 = 56$ $56 \div 7 = 8$ $56 \div 8 = 7$ b 21 $3 \times 7 = 21$ $7 \times 3 = 21$ $21 \div 7 = 3$ $21 \div 3 = 7$	**On track** • Can recall multiplication facts or use an appropriate method (e.g. known facts, commutativity, splitting factors) to work out facts they cannot instantly recall. • Can use knowledge of the link between multiplication and division to write families of four related facts. **Further learning required** • May not have recall of multiplication/division facts and rely on less efficient strategies such as sharing and grouping concrete materials. • May not understand the relationships between multiplication and division (commutativity and/or inverse) and so be unable to write two multiplication and two division number sentences to match each triangle.
MTH 2-03a	26	Explanations of the links between the 2, 4 and 8 times tables will vary. There should be some mention of 'doubles' and/or 'halves'.	**On track** • Can confidently link the 8, 4 and the 2 times tables to 'doubles' and 'halves' and give examples. **Further learning required** • May 'see' the links between the 8, 4 and the 2 times tables but be unable to explain them.
MNU 2-02a MNU 2-03b	27	a True b False: $1000 \times 3.56 = 3560$ c False: $0.04 \times 10 = 0.4$ d False: $267.37 \times 100 = 26\,737$	**On track** • Can confidently use place value understanding to multiply decimal fractions by 10, 100 and 1000. **Further learning required** • May be unable to identify the mistakes in the number sentences. • May multiply by 10, 100 and/or 1000 incorrectly. • May be able to identify b and c as false but for the wrong reason, e.g. believe you 'add a zero' when multiplying by 10.

<table>
<tr><th>Topic – benchmarks / Es & Os</th><th>Question</th><th>Answer</th><th>Marking guidance</th></tr>
<tr><td>MNU 2-02a
MNU 2-03b</td><td>28</td><td>Isla started with 2800.</td><td>On track<ul><li>Understands the multiplicative relationship between the places in decimal fractions, i.e. 10/100/1000 bigger/smaller.</li><li>Understands that halving is the inverse of doubling.</li><li>Uses knowledge of multiplication as the inverse of division to solve the problem, i.e. works backwards, multiplying by 10, then halving, then multiplying by 1000 to find the starting number.</li></ul>Further learning required<ul><li>May not understand the multiplicative relationship between the places and tries to apply mathematically incorrect rules such 'Remove a zero when dividing by 10' to decimal numbers.</li><li>May not understand the inverse relationship between multiplication and division/ doubling and halving and so be unable to work backwards to find the starting number.</li></ul></td></tr>
<tr><td>MNU 2-03a</td><td>29</td><td>a 4186 × 6 = 25 116<table><tr><td>×</td><td>4000</td><td>100</td><td>80</td><td>6</td></tr><tr><td>6</td><td>24 000</td><td>600</td><td>480</td><td>36</td></tr><tr><td colspan="5">24 000 + 600 + 480 + 36 = 25 116</td></tr></table>
b 5651 × 5 = 28 255<table><tr><td>×</td><td>5000</td><td>600</td><td>50</td><td>1</td></tr><tr><td>5</td><td>25 000</td><td>3000</td><td>250</td><td>5</td></tr><tr><td colspan="5">25 000 + 3000 + 250 + 5 = 28 255</td></tr></table>
c 3294 × 7 =<table><tr><td>×</td><td>3000</td><td>200</td><td>90</td><td>4</td></tr><tr><td>7</td><td>21 000</td><td>1400</td><td>630</td><td>28</td></tr><tr><td colspan="5">21 000 + 1400 + 630 + 28 = 23 058</td></tr></table></td><td>On track<ul><li>Understands the concept of using the grid method for multiplying.</li><li>Can partition four-digit numbers into thousands, hundreds, tens and ones to support the grid method.</li><li>Understands the multiplicative relationship between the place value places in order to multiply by increasing powers of 10.</li><li>Can confidently use known multiplication facts to support multiplying by thousands and hundreds.</li><li>Accurately adds the individual totals to calculate the overall answer.</li></ul>Further learning required<ul><li>May be unfamiliar with the grid method for multiplying.</li><li>May not be able to use place value partitioning to thousands.</li><li>May not have recall of multiplication/ division facts and rely on less efficient strategies such as sharing and grouping concrete materials.</li></ul></td></tr>
</table>

Topic – benchmarks / Es & Os	Question	Answer	Marking guidance
			• May not understand the multiplicative relationship between the places. • May be unable to calculate the overall total, e.g. part b may be easier as the total does not involve bridging over a thousand.
MNU 2-03b	30	a £5.37 b £11.75 c £28.02	**On track** • Identifies multiplication as the required operation. • Can confidently use a variety of strategies to multiply decimal fractions by a single digit number, e.g. using a grid method or round and compensate. • Chooses an efficient strategy appropriate to the numbers in the calculation. **Further learning required** • May not identify the problems as multiplication. • May calculate inaccurately; possibly due to incorrect recording of the numbers in the calculation, the strategy chosen or choosing an appropriate strategy but using it ineffectively. • May be confused when working with decimal fractions to two places, e.g. may answer £28.2 or £28.20 for c. • May be over reliant on one strategy that may be inefficient.
MNU 2-03a	31	a $552 \div 6 = 92$ b $702 \div 9 = 78$ c $344 \div 8 = 43$	• Can confidently use known facts and place value partitioning, or a different strategy of their choice, to divide three-digit whole numbers by a single digit number. **Further learning required** • May calculate inaccurately; possibly due to incorrect recording of the numbers in the calculation, the strategy chosen or choosing an appropriate strategy but using it ineffectively, e.g. choosing partitioning but partitioning the number incorrectly. • May be over reliant on only one strategy that may not be the most efficient.

Topic – benchmarks / Es & Os	Question	Answer	Marking guidance
MNU 2-03a	32	a Amman earns £126 in seven weeks. b Finlay baked 60 cakes.	**On track** • Identifies multiplication as the required operation. • Calculates accurately using mental strategies such as compensation, known facts and partitioning and/or informal jottings. • May choose different strategies, depending on the calculation. **Further learning required** • May not identify the problems as multiplication, or chooses to use an inefficient method such as repeated addition with concrete materials, e.g. 18 + 18 + 18 + 18 + 18 + 18 + 18 instead of mental partitioning and jottings (7 x 10) + (7 x 8). • May calculate inaccurately; possibly due to incorrect recording of the numbers in the calculation, the strategy chosen or choosing an appropriate strategy but using it ineffectively. • May be over reliant on one strategy that may not be the most efficient.
MNU 2-03a	33	a 23 tables can be made – if there were 100 legs, 25 tables could be made, therefore 92 legs can make 23 tables. b 41 bags are required – if there were 160 bulbs, 40 bags would be required, therefore 41 bags will be needed for 164.	**On track** • Can interpret the problem as being about equal groups and make links with multiplication and division. • Understands the link between multiplication and division and can use their knowledge of division as the inverse of multiplication to solve the problem. • Can use rounding and compensating to find solutions. **Further learning required** • May be unable to visualise or represent the problems or does so incorrectly, e.g. may represent the problem as multiplication not division. • May be able to select the appropriate operation but need support to recall the multiplication/division facts required to complete the calculations, relying on concrete materials, drawings or diagrams to work each answer out. • May be unable to use rounding and compensation to find solutions, e.g. may divide 100 by 4 but forget to compensate by taking the '8 extra legs' away.

Topic – benchmarks / Es & Os	Question	Answer	Marking guidance
MNU 2-03a	34	a $14 \times 23 = 322$ b $16 \times 28 = 448$ c $7 \times 84 = 588$	**On track** • Calculates accurately. Attention should also be paid to the strategies used, e.g. using grid partitioning, compensation or algorithm. • May choose different strategies, depending on the calculation. **Further learning required** • May calculate inaccurately; possibly due to incorrect recording of the calculation, the strategy chosen or choosing an appropriate strategy but using it ineffectively, e.g. multiplying by 30 to solve b but then not compensating by subtracting 32 (16×2). • May be reliant on one strategy only that may not be the most efficient.
MNU 2-02a MNU 2-03a	35	a $135 \div 5 = 27$ b $342 \div 8 = 42.75$ c $112 \div 8 = 14$ d $2.64 \div 4 = 0.66$ e $2.88 \div 3 = 0.96$ f $4.41 \div 7 = 0.63$	**On track** • Calculates accurately. Attention should be paid to the strategies used, e.g. place value partitioning or an algorithm. • When using algorithms, can express remainders as decimal fractions. • Understands the importance of zero as a place holder in d, e and f. **Further learning required** • May calculate inaccurately; possibly due to incorrect recording/ layout of the calculation, the strategy chosen or choosing an appropriate strategy but using it ineffectively. • May choose to leave the remainder as a whole number instead of converting it to a decimal fraction. • May not understand the importance of zero as a place holder.

Topic – benchmarks / Es & Os	Question	Answer	Marking guidance
MNU 2-03a MTH 2-03c	36	a Isla will have £26 left £18 + £36 = £54 £80 – £54 = £26 b No, Finlay will get the wrong answer. The order of operations is important when there are addition, subtraction, multiplication and division calculations in problems. The correct answer is $6 + (8 \div 2) = 10$.	**On track** • Chooses the correct operations to solve the problem. • Calculates accurately. Attention should be paid to the strategies used, how numbers are partitioned and/or the methods chosen to potentially make calculating easier. • Understands the order of operations and their importance in problems. **Further learning required** • May use an inefficient strategy such as repeated subtraction, i.e. 80 – 9 – 9 – 12 – 12 –12. • May calculate inaccurately; possibly due to incorrect recording of the calculation, the strategy chosen or choosing an appropriate strategy but using it ineffectively. • May not understand the importance of the order of operations.
Multiples, factors and primes MTH 2-05a	37	725, 1345 are divisible by 5. 150, 2770 and 1010 are divisible by 5 and 10.	**On track** • Understands the divisibility rules for dividing by 5 and 10. **Further learning required** • May be confused about divisibility rules for dividing by 5 and 10, e.g. may think the digits should add to 5 or 10.
MTH 2-05a	38	<table><tr><th>Problem</th><th>Isla's answer</th><th>Finlay's answer</th></tr><tr><td>27×10</td><td>**270**</td><td>271</td></tr><tr><td>$135 \div 2$</td><td>65</td><td>**67.5**</td></tr><tr><td>26×9</td><td>**234**</td><td>270</td></tr></table>Explanations may vary but could mention rules about multiples and factors or numbers being too large to be the answer.	**On track** • Uses knowledge of multiples and factors to solve problems. **Further learning required** • May be unable to explain why an answer is correct/incorrect without completing the calculation to check, e.g. may not be able to explain why the answer to $135 \div 2$ cannot be a whole number.

Topic – benchmarks / Es & Os	Question	Answer	Marking guidance
Fractions, decimal fractions and percentages MTH 2-07c	39	$\frac{2}{8}, \frac{3}{10}, \frac{2}{5}, \frac{1}{2}, \frac{3}{4}$.	**On track** • Can identify the common factor (40), convert each fraction into fortieths and use this information to order the fractions listed from smallest to largest. **Further learning required** • May be unable to find a common denominator in order to compare fractions. • May not understand how the denominator indicates the size of the whole represented and be unable to order the fractions from smallest to largest, or orders them incorrectly. • May add the numerator and denominator together in order to compare size.
MTH 2-07c	40	a $\frac{35}{60} = \frac{7}{12}$ b $\frac{60}{100} = \frac{3}{5}$ c $\frac{9}{36} = \frac{1}{4}$ Diagrams are drawn on squared paper to prove the simplified fractions, e.g. that $\frac{9}{36}$ is the same as $\frac{1}{4}$.	**On track** • Understands how to simplify fractions by identifying common factors. • Understands the phrase 'simplest form' and can write fractions in their simplest form. • Can prove their solutions by drawing diagrams on squared paper. **Further learning required** • May require support to identify common factors. • May simplify each fraction (e.g. simplifies $\frac{60}{100}$ to $\frac{6}{10}$) but be unable to express them in their simplest form. • May be unable to draw accurate diagrams to prove they have recorded the fraction in its simplest form, e.g. may use two different-sized wholes to compare the same fraction.

Topic – benchmarks / Es & Os	Question	Answer	Marking guidance
MNU 2-07a	41	Isla travelled $4\frac{1}{4}$ miles.	**On track** • Can visualise the problem and represent the information given as a bar model or diagram. • Can use mental strategies, supported by informal jottings, to calculate with wholes and fractions. **Further learning required** • May be unable to visualise and represent the problems. • May be confused when bridging a whole, e.g. when adding $2\frac{1}{2}$ and $3\frac{3}{4}$.
MNU 2-07b MTH 2-07c	42	a $0.75 = \frac{75}{100} = \frac{3}{4} = 75\%$ b $0.4 = \frac{40}{100} = \frac{2}{5} = 40\%$ c $0.15 = \frac{15}{100} = \frac{3}{20} = 15\%$	**On track** • Can equate each decimal number with hundredths, i.e. knows that 0.75 means $\frac{75}{100}$; 0.4 means $\frac{40}{100}$; $0.15 = \frac{15}{100}$ • Can write these fractions in their simplest form. • Understands that percent means 'out of 100' and use this knowledge to convert each fraction to a percentage. • Can equate 0.4 (four tenths) with 0.40 (forty hundredths). **Further learning required** • May be able to read decimal fractions as 'zero point...' but be unable to rewrite the decimal fractions as fractions and/or as percentages. • May not know that percent means 'out of 100' and therefore cannot rewrite the decimal fraction as a percentage (may have memorised 0.75 as 75%). • May record 0.4 as 4%. • May not understand the phrase 'simplest form'. • May attempt to simplify the fractions but make division errors.

Topic – benchmarks / Es & Os	Question	Answer	Marking guidance
MNU 2-07a	43	a Finlay has $\frac{1}{6}$ of the bag of sweets. b Isla has $\frac{1}{6}$ of the whole bottle of water left.	**On track** • Can visualise and represent each problem pictorially or as a diagram, for example as a bar model. • Can use the representations created to support their thinking and solutions. **Further learning required** • May be unable to visualise and/or represent the word problems. • May be unable to visualise/represent the whole, the fractional part that Isla starts with and the fractional part given away. • May be unable to complete the calculations, e.g. cannot work out a third of a half, even when supported by drawings or diagrams .
MNU 2-07a	44	Amman can make 28 scones. Methods/drawings and calculations to find the answer may vary.	**On track** • Can understand and visualise the word problem. • Can represent the information given in the word problem, possibly by drawing a picture or diagram such as a bar model, or by using known facts supported by informal jottings. **Further learning required** • May be unable to visualise or represent the word problem. • May not equate 3 wholes with 24 eighths and/or $\frac{1}{2}$ with $\frac{4}{8}$ • May omit parts of the calculation and/or make calculation errors, e.g. calculates the number of scones by multiplying 3 by 8 then adding 8.
MNU 2-07a	45	Isla will have 40 portions of chocolate. Methods/drawings and calculations to find the answer may vary.	**On track** • Can understand and visualise the word problem. • Can represent the information given in the word problem, possibly by drawing a picture or diagram such as a bar model or using known facts supported by informal jottings. **Further learning required** • May be unable to visualise or represent the word problem. • May not equate 5 wholes with 40 eighths.

Topic – benchmarks / Es & Os	Question	Answer	Marking guidance
MNU 2-07b	46	15% of 6500 is 975. Working out may include: 10% of 6500 = 650 5% of 6500 = 325	**On track** • Can understand and visualise the word problem. • Can represent the information given in the word problem. • Knows that to calculate 10% they should divide by 10 and can use this knowledge to work out 5%. **Further learning required** • May be unable to visualise or represent the word problem. • May be unable to calculate a percentage of an amount. • May omit parts of the calculation or calculate the percentage incorrectly.
Money MNU 2-09a	47	Nuria's mum should buy a book of four return tickets (£8.50) plus one return ticket (£2.25) for her week at work. Total cost £10.75	**On track** • Can understand and visualise the word problem. • Can represent the information given in the word problem, choose the correct operations and uses mental maths strategies, supported by informal jottings, to calculate the difference between the costs of various ticket options. • Can state which ticket option Nuria's mum should choose, justifying their thinking. **Further learning required** • May be unable to visualise the word problem and/or extract the correct information. • May choose the wrong operations. • May make calculation errors. • May be unable to show/explain their reasoning.

Topic – benchmarks / Es & Os	Question	Answer	Marking guidance
MNU 2-09a MNU 2-09b	48	Finlay's gran has £540 per month left for other essentials. £450 + £160 = £610 £1150 – £610 = £540	**On track** • Can understand and visualise the word problem. • Can represent the information given in the word problem, choose the correct operations and use mental maths strategies, supported by informal jottings, to calculate how much money gran has left. **Further learning required** • May be unable to visualise the word problem and/or extract the correct information. • May choose the wrong operations. • May make calculation errors. • May be unable to show/explain their reasoning.
MNU 2-09c	49	Amman and his dad made a profit of £32.50. Sofa = £50 Fabric = £25.50 Materials = £12 Total = £87.50 £120 – £87.50 = £32.50	**On track** • Understands what 'profit' means. • Can interpret the word problem. • Can represent the information given in the word problem, choose the correct operations and use mental maths strategies, supported by informal jottings, to calculate the profit made. **Further learning required** • May be unable to visualise the word problem and/or extract the correct information. • May choose the wrong operations. • May make calculation errors. • May be unable to show/explain their reasoning. • May not understand the term 'profit'.

Topic – benchmarks / Es & Os	Question	Answer	Marking guidance
MNU 2-09a	50	Isla should take the £5 discount offer to reduce the cost of the flowers to £25. A 15% discount would reduce the cost to £25.50.	**On track** • Can understand and visualise the word problem. • Can represent the information given in the word problem, choose the correct operations and use mental maths strategies, supported by informal jottings, to calculate which is the better offer. • Understands the term 'discount'. • Can calculate with percentages. **Further learning required** • May be unable to visualise the word problem and/or extract the correct information. • Does not understand the term 'discount'. • May choose the wrong operations. • May make calculation errors. • May be unable to show/explain their reasoning. • May be unable to accurately calculate 15%. • May be unable to state which would be the better offer.
MNU 2-09b	51	Hire purchase cost: 12 x £35 = £420 Extra paid = £420 – £240 = £180	**On track** • Understands the phrase 'hire purchase agreement'. • Can interpret the word problem and carry out the appropriate calculations to solve it, i.e. - knows to multiply the monthly fee by 12 months. - knows to find the difference between this and the cash price. **Further learning required** • May not understand and/or be unable to visualise the word problem. • May not know how hire purchase works. • May choose the wrong operations. • May make calculation errors. • May be unable to show/explain their reasoning. • May not remember how many months there are in a year. • May fail to complete all the steps in the problem.

Topic – benchmarks / Es & Os	Question	Answer	Marking guidance
Time MNU 2-10a	52	Finlay arrives at 1715.	**On track** • Can read and write 24-hour time. • Can calculate the arrival time by adding the duration 3 h 50 min to the start time, possibly using a number line and informal jottings. • Can calculate duration when bridging over hours. **Further learning required** • May have difficulty reading and calculating 24-hour time, particularly when working with examples that bridge an hour. • May choose an inappropriate strategy or use the strategy chosen incorrectly, e.g. may attempt to use a formal written algorithm instead of mental strategies, supported by an empty number line and/or informal jottings.
MNU 2-10a	53	The duration of the competition is 4 days and 8 hours.	**On track** • Can understand 12-hour times and a.m./p.m. notation. • Can use 12-hour time to calculate duration in days and hours. **Further learning required** • May confuse 12-hour and 24-hour time notation. • May calculate incorrectly, possibly because they do not realise that they can add on days, then hours and so tries to count on in hours only. • May count on incorrectly, i.e. includes Monday in the total number of days.
MNU 2-10c	54	a The cyclist takes 4 hours to travel 80 km. b The car travels 150 miles in 3 hours.	**On track** • Can visualise and correctly interpret each word problem. • Chooses the correct formula for each problem and uses it correctly. • Uses knowledge of multiplication and division to accurately solve the problems. **Further learning required** • May be unable to correctly use the time/distance/speed formulae or be confused by the context of the word problem. • May be unable to multiply and divide by multiples of 10.

<table>
<tr><th>Topic – benchmarks / Es & Os</th><th>Question</th><th>Answer</th><th>Marking guidance</th></tr>
<tr><td>MNU 2-10b</td><td>55</td><td><table><tr><td>Start time</td><td>Finish time</td><td>Time elapsed (duration)</td></tr><tr><td>3 min 24 s</td><td>4 min 15 s</td><td>51 s</td></tr><tr><td>5 min 46 s</td><td>10 min 24 s</td><td>4 min 38 s</td></tr><tr><td>12 min 15 s</td><td>16 min 5 s</td><td>3 min 50 s</td></tr></table></td><td>On track
• Can read and interpret the table.
• Can choose between counting on and counting back / addition and subtraction, to complete the table.
• Can calculate duration in minutes and seconds, including examples that bridge a minute, using mental calculation strategies, supported by an empty number line and/or informal jottings.
Further learning required
• May have difficulty interpreting the table.
• May choose the wrong operation for some or all of the examples.
• May be unable to choose and use an appropriate strategy, e.g. may attempt to incorrectly use a formal written algorithm instead of counting on or back on an empty number line.</td></tr>
<tr><td>Measurement
MNU 2-11b</td><td>56</td><td>a 458 cm = 4.58 m
b 1.4 km = 1400 m
c 3.4 cm = 34 mm
d 5310 m = 5.31 km</td><td>On track
• Knows equivalences such as 10 mm = 1 cm, 100 cm = 1 m and 1000 m = 1 km
• Understands the multiplicative relationship between the places in our number system and can use this knowledge to convert between units of measurements.
Further learning required
• May not know that 1 cm = 10 mm,100 cm = 1 m and 1000 m = 1 km.
• May be able to equate, for example, 458 cm with 4 m and 58 cm but be unable to relate this to the decimal form 4.58 m.
• May be able to convert smaller units to larger units (e.g. centimetres to metres) but find converting larger units to smaller units more challenging.
• Lack of place value understanding may impact upon calculations, e.g. writes 53.10 km for d.</td></tr>
</table>

Topic – benchmarks / Es & Os	Question	Answer	Marking guidance
MNU 2-11b	57	Amman's bag weighs 4.25 kg/4250 g/4 kg 250 g on the way home from the shop. Carrots = 1.4 kg Potatoes = 2.1 kg Broccoli = 750 g/0.75 kg	**On track** • Can use their knowledge of kilograms and place value to calculate the value of each graduation on the scales, i.e. 0.1 kg/100 g. • Reads the scales accurately and can work out the mass of each object using the correct unit of measurement, i.e. grams or kilograms. • Uses the scale measurements to accurately calculate the total mass using a method of their choice. **Further learning required** • May not understand, or know how to calculate, the value of each graduation on the scales and so is unable to record the mass of each object correctly. • May confuse grams and kilograms. • May calculate incorrectly, e.g. add 750 to 1.4 and 2.1.
MNU 2-11c	58	The missing lengths are: Mobile phone: 10 cm Tablet: 22 cm Garden: 6.3 m	**On track** • Can use their knowledge of the properties of a rectangle to calculate the length of the unknown sides when given the area and length of one side. • Understands division as the inverse of multiplication and can combine this with their knowledge of the formula area = length × breadth to calculate the length of the unknown side. • Uses the correct units to record the length of the unknown side, i.e. cm not cm^2. **Further learning required** • May not know the formula for finding the area of a rectangle. • May know the formula but be unable to use the inverse to find the length of the unknown side. • May be unable to choose and/or use an appropriate method for calculations. • May use incorrect units of measurement to record the length of the unknown side.

<table>
<tr><th>Topic – benchmarks / Es & Os</th><th>Question</th><th>Answer</th><th>Marking guidance</th></tr>
<tr><td>MNU 2-11b</td><td>59</td><td>a 180 ml or 0.18 l
b 260 ml or 0.26 l
c 720 ml or 0.72 l</td><td>On track<ul><li>Can use their knowledge of millilitres, litres and place value to calculate the value of each graduation on the scales, i.e. 0.1 l /100 ml.</li><li>Can read the scales accurately and can work out each measurement using the correct unit, i.e. millilitres and litres.</li></ul>Further learning required<ul><li>May not understand or know how to calculate the value of each graduation on the scales and so is unable to record the volume of liquid in each jug correctly.</li><li>May confuse millilitres and litres.</li><li>May not equate 100 ml with 0.1 l.</li></ul></td></tr>
<tr><td>MNU 2-11a</td><td>60</td><td><table>
<tr><td>Item being measured</td><td>Metric units that could be used</td><td>Imperial units that might be used</td></tr>
<tr><td>The distance between Edinburgh and Aberdeen</td><td>km</td><td>miles</td></tr>
<tr><td>The height of an adult</td><td>metres</td><td>Feet and inches</td></tr>
<tr><td>The mass of an adult</td><td>kg</td><td>Stones and pounds</td></tr>
<tr><td>The capacity of a container of milk</td><td>litres</td><td>pints</td></tr>
</table></td><td>On track<ul><li>Can understand the difference between metric and imperial units.</li><li>Can understand the vocabulary used to measure mass, volume and distance in metric and imperial measurements.</li></ul>Further learning required<ul><li>May confuse metric and imperial measurement vocabulary.</li></ul></td></tr>
</table>

Topic – benchmarks / Es & Os	Question	Answer	Marking guidance
MNU 2-11c	61	Answers will vary. The three zones of the playground should have perimeters of 50 m, 32 m and 18 m. The area of each zone should be calculated correctly.	**On track** • Can follow detailed instructions and keep track of their progress through the question. • Understands that perimeter means 'all the way round the edge'. • Can use reasoning skills to identify possible 2D shapes that fit the criteria set. • Can add accurately to calculate perimeter and record measurements in metres (m). • Is familiar with squares, rectangles and composite shapes. • Can multiply accurately to calculate area and record in square metres (m^2). • Knows they can split composite shapes into simpler shapes that are easier to calculate with. **Further learning required** • May be unfamiliar with the words perimeter and area, or confuse perimeter and area. • May be unsure of the properties of 2D shapes and use a 'trial and error' approach, measuring each side individually. • May struggle to multiply some of the measurements or be unsure about effective strategies to calculate the perimeter, especially for calculations involving 0.5 cm. • May be unable to calculate using the measurements or be unsure about effective strategies to calculate the area, e.g. by dividing the composite shapes into simple shapes. • May not use cm^2 as a unit of measurement.

Topic – benchmarks / Es & Os	Question	Answer	Marking guidance
MNU 2-11c	62	a $(6 \times 12) \div 2 = 36\ cm^2$ b $(6 \times 6) \div 2 = 18\ cm^2$ c $(8 \times 14) \div 2 = 56\ cm^2$	**On track** • Can use their knowledge of the properties of right-angled triangles to find the surface area of each, e.g. knows that a right-angled triangle is half of a square/rectangle or may apply a formula for the area. • Can multiply and divide accurately to calculate surface area and use cm^2 to record this. **Further learning required** • May be unfamiliar with the word area or confuse area and perimeter. • May be unsure of the properties of right-angled triangles and so rely on measuring each side with a ruler and stating the perimeter instead of the area. • May struggle to multiply/divide some of the measurements. • May multiply length × breadth but not realise they have to halve their answers.
Mathematics, its impact on the world, past, present and future MTH 2-12a	63	Babylonian symbols to represent the numbers 1 and 19.	**On track** • Can use the Babylonian number system to create the numbers 1 and 19. **Further learning required** • May not know what a factor is.
Patterns and relationships MTH 2-13a	64	a 1, 3, 6, 10, 15, 21 – triangular numbers b 1, 4, 9, 16, 25, 36 – square numbers Diagrams should be drawn to prove whether patterns contain triangular or square numbers.	**On track** • Can identify, continue and explain each sequence. • Understands the difference between square and triangular numbers. • Uses diagrams to demonstrate square and triangular numbers. **Further learning required** • May have difficulty identifying and continuing each sequence. • May be unfamiliar with triangular and square numbers so be unable to associate each sequence with this language and/or draw diagrams to prove their answers.

Topic – benchmarks / Es & Os	Question	Answer	Marking guidance
Expressions and equations MTH 2-15a	65	a Triangle = 3 b Circle = 8	**On track** • Understands the concept of balance in equations. • Can reason that the same shape on both sides of the balance cancel each other out. • Can identify similarities and differences to work out the numerical values the shapes represent. **Further learning required** • May not understand the concept of balance in equations, e.g. may not realise that the squares on each side of the balance cancel each other out. • May be unable to work out the numerical values the shapes represent.
2D shapes and 3D objects MTH 2-16c	66	Can draw: a a scalene triangle with one side 8 cm and one angle of 60° b an isosceles triangle with a base that is 6 cm long c an equilateral triangle.	**On track** • Knows that the angles of a triangle add to 180 degrees. • Understands the difference between scalene, isosceles and equilateral triangles and can identify their properties. • Can competently use the protractor to draw given angles. • Can use a ruler to draw sides of a given length. • Coordinates both of these steps to produce triangles of a given type. **Further learning required** • May confuse the properties of scalene, isosceles and equilateral triangles. • May know they must create given angles but cannot work out how to use a protractor to do this. • May have difficulty using a ruler in coordination with the protractor to draw the triangles, e.g. may just use the ruler and estimate the angles.
MTH 2-16c	67	Recognisable sketches of each shape inserted into the correct place in the table. Three-sided shapes: equilateral triangle, isosceles triangle Four-sided shapes: rectangle, square, parallelogram, trapezium	**On track** • Can visualise all of the 2D shapes named, understands their properties and can make realistic sketches of these in different orientations, i.e. so they do not all have a horizontal base line. **Further learning required** • May be unable to recall the names or properties of the 2D shapes mentioned. • May struggle to visualise some shapes or have difficulty drawing them so they are not orientated along a horizontal line.

Topic – benchmarks / Es & Os	Question	Answer	Marking guidance
MTH 2-16a	68	a **B** b **B** c **A**, **B**, **C** and **D** d **A** and **D**	**On track** • Can understand the vocabulary related to 3D objects, i.e. face, vertex. • Can visualise the 3D objects represented by the skeletal models and sort them according to the properties given. **Further learning required** • May be unfamiliar with the vocabulary relating to 3D objects and so be unable to identify or sort the skeletal models according to given properties. • May have difficulty visualising 3D objects from 2D drawings.
MTH 2-16b	69	a cone b triangular prism c square-based pyramid d triangular prism	**On track** • Can identify and name each 3D object from its net. **Further learning required** • May be unable to match the net to the correct 3D object. • May be able to match each net to the correct solid but be unable to name them.
Angles, symmetry and transformation MTH 2-17b	70	a Side AC = 5.4 cm Side BC = 4.4 cm Angle ABC = 66° Angle ACB = 66° b Side AB = 5.6 cm Side BC = 4.2 cm Angle ACB = 43° Angle BAC = 30° (note that these answers could differ slightly depending on whether the page has been scaled when printing/ photocopying)	**On track** • Can recall the properties of triangles. • Can measure angles accurately using a protractor and record these in degrees. • Can measure lengths accurately using a ruler and record these in centimetres. **Further learning required** • May be unable to use a protractor or ruler accurately. • May be unable to use strategies to help them measure angles or lengths of sides, e.g. rotating the paper to measure the angle from a different perspective. • May be unable to record measurements using the appropriate unit, or records incorrectly.

Topic – benchmarks / Es & Os	Question	Answer	Marking guidance
MTH 2-17a	71	a 168 degrees b 95 degrees c 15 degrees	**On track** • Understands the vocabulary acute, obtuse and reflex. • Knows that a 'full turn' equals 360 degrees. • Can accurately calculate acute or obtuse angles by subtracting the reflex angle from 360 degrees. **Further learning required** • May not understand the vocabulary or properties of acute, obtuse and reflex angles. • May be unable to calculate acute or obtuse angles accurately, e.g. may subtract from a number other than 360 degrees or may make mistakes in calculation.
MTH 2-17a	72	Angle A: 150 degrees Angle B: 65 degrees Angle C: 42 degrees Angle D: 42 degrees Angle E: 228 degrees	**On track** • Can use known properties of right angles, full turns, and straight angles to calculate complementary and supplementary angles accurately. **Further learning required** • May not know the properties of right, full turns or straight angles. • May confuse the complementary and/or supplementary angles. • May be unable to complete the calculations accurately.
MTH 2-17c	73	Nuria is now facing north-west.	**On track** • Can relate the eight-point compass with angles measured in degrees, i.e. knows the angle between each point measures 45 degrees. • Can calculate the angle between different points on an eight-point compass. **Further learning required** • May be unable to understand the relationship between points of an eight-point compass and the angles between the points. • May make mistakes when calculating the angle between different points on an eight-point compass.

Marking guidance

<table>
<tr><th>Topic – benchmarks / Es & Os</th><th>Question</th><th>Answer</th><th>Marking guidance</th></tr>
<tr><td>MTH 2-18a</td><td>74</td><td>D (3,13), E (7,13)</td><td>On track
• Understands that congruent means 'the same shape and size' and can use this knowledge to reason about the problem and calculate the missing vertices.
• Can accurately record the coordinates of the missing vertices.
Further learning required
• May not know what 'congruent' means and hence be unable to find the correct coordinates or even start the problem.
• May be unable to write the coordinates correctly, e.g. writes (1,0) as (0.1).</td></tr>
<tr><td>MTH 2-19a</td><td>75</td><td>a
<table><tr><td></td><td>One or more diagonal line of symmetry</td><td>Vertical and horizontal lines of symmetry</td></tr><tr><td>5 sides or more</td><td>B</td><td>D, E</td></tr><tr><td>5 sides or fewer</td><td>A, C</td><td></td></tr></table>b F does not belong.</td><td>On track
• Can identify the lines of symmetry in the given shapes.
• Understands the difference between diagonal, horizontal and vertical symmetry.
• Understands how to use a Carroll diagram.
Further learning required
• May be unable to identify lines of symmetry on the shapes given, e.g. may believe the shape F has horizontal and vertical lines of symmetry.
• May be unable to complete the Carroll diagram.</td></tr>
<tr><td>MTH 2-19a</td><td>76</td><td>Design created with 4 lines of symmetry.</td><td>On track
• Can identify and work with diagonal, horizontal and vertical lines of symmetry.
• Can accurately reflect the pattern in all lines of symmetry.
• The pupil may add some of their own shapes when creating the symmetrical pattern.
Further learning required
• May be unable to identify lines of symmetry.
• May complete the design inaccurately, e.g. may only use one line of symmetry to complete the design or may simply transfer the design onto each segment.</td></tr>
</table>

Topic – benchmarks / Es & Os	Question	Answer	Marking guidance
MTH 2-17d	77	Lines of the following lengths drawn and labelled: a Edinburgh Castle – Kelpies: 4.5 cm b Kelpies – Wallace Monument: 3 cm c Culloden – Dunrobin Castle: 9 cm d John O'Groats – Ullapool: 19 cm	**On track** • Understands the relationship between the scale given and how to calculate this as a length to represent distance. • Can divide whole numbers by 10, creating decimal numbers. • Can confidently use a ruler, using cm and mm, to draw the distance between the locations to scale. • Can use effective mental maths strategies or informal jottings to calculate distances. **Further learning required** • May not know how to use the scale to create the scale lengths. • May be unable to use a ruler correctly to measure the distance between places and their scale distances. • May have difficulty calculating the actual distance between places when dividing by 10.
Data handling and analysis MNU 2-20b	78	a 1) The survey was only conducted on a Friday and 2) Only the Primary 7 class was surveyed. b A more reliable sample would be to ask every class in the school on the same day about their thoughts on after-school clubs.	**On track** • Can use knowledge of sampling to evaluate the validity of the information given. • Can suggest how the survey could be improved. **Further learning required** • May not understand the concept of bias. • May be unable to suggest how to improve the survey.

Topic – benchmarks / Es & Os	Question	Answer	Marking guidance
MNU 2-20a	79	Answers may vary. Responses may include: Gym customers: The graph clearly shows when the gym is busiest. A scale of 10 on the y-axis shows small changes in the number of customers. Newspapers: The bar graphs make a poor comparison between two newspaper sales. Beginning the y-axis at 400 000 and having a scale of 25 000 distorts the data in favour of the Herald. Sales: This graph highlights the constant increase in sales from April to August. Using a scale of 1000 and extending this to 6000 may show where the company hope the sales will go or have been in the past.	**On track** • Can confidently read and interpret line and bar graphs. • Understands how changes in scale can influence the way a graph is interpreted. • Can suggest reasons for presenting information in specific ways, considering the author, audience and purpose. **Further learning required** • May be unable to read and interpret the line and/or bar graphs. • May be unable to identify how the changes in scale affect the information presented in the graph. • May be unable to identify reasons and/or explain why graphs may be drawn in a certain way.

<table>
<tr><th>Topic – benchmarks / Es & Os</th><th>Question</th><th>Answer</th><th>Marking guidance</th></tr>
<tr><td>MTH 2-21a</td><td>80</td><td>a
<table><tr><td>Product</td><td>Number of degrees</td><td>Percentage of the total</td></tr><tr><td>Porridge</td><td>180</td><td>50%</td></tr><tr><td>Melon</td><td>108</td><td>30%</td></tr><tr><td>Apple</td><td>72</td><td>20%</td></tr><tr><td>TOTAL:</td><td>360</td><td>100%</td></tr></table>b Pie chart completed and labelled correctly
c The three products total 360 degrees and 100%. Porridge accounted for half the total sales.</td><td>On track
• Understands that a pie chart has a total of 360° that represents 100% or the whole pie chart.
• Reasons that a pie chart is composed of fractional parts and uses this knowledge to calculate the number of degrees for each part.
• Uses knowledge of the link between fractions and percentages to interpret the table and calculate percentages.
• Can use a protractor to measure degrees to draw the pie chart.
• Makes links between the percentages and the degrees in each section of the pie chart.
• Labels the pie chart correctly.
Further learning required
• May recognise that porridge accounted for half of the products sold but does not make the link with percentages.
• May not understand how to divide the pie chart into fractional parts.
• Does not know the relationship between fractions and percentages or that the whole represents 100%.
• May not know how to start drawing the angles or may not use the protractor correctly, e.g. not starting the measurement at zero degrees.
• May create a pie chart but not label the parts.</td></tr>
<tr><td>Ideas of chance and uncertainty
MNU 2-22a</td><td>81</td><td>a $\frac{1}{2}$; 50%, even chance
b 12 times. You have a one in six chance of rolling a 3 because the dice has six sides. $72 \div 6 = 12$.</td><td>On track
• Can reason about the likelihood of chance and probability events.
Further learning required
• May be unable to interpret the problem.
• May understand the problem but be unable to carry out the calculations necessary to solve it.</td></tr>
</table>

End of second level assessment 1

The activity tracker

Curriculum organisers	Experiences and outcomes	Benchmarks
Estimation and rounding	I can use my knowledge of rounding to routinely estimate the answer to a problem, then, after calculating, decide if my answer is reasonable, sharing my solution with others. MNU 2-01a	• Rounds whole numbers to the nearest 1000, 10 000 and 100 000. • Applies knowledge of rounding to give an estimate to a calculation appropriate to the context.
Number and number processes	Having determined which calculations are needed, I can solve problems involving whole numbers using a range of methods, sharing my approaches and solutions with others. MNU 2-03a I have explored the contexts in which problems involving decimal fractions occur and can solve related problems using a variety of methods. MNU 2-03b	• Adds and subtracts whole numbers and decimal fractions to two decimal places, within the number range 0 to 1 000 000. • Multiplies and divides whole numbers by multiples of 10, 100, 1000. • Divides whole numbers and decimal fractions to two decimal places, by a single digit, including answers expressed as decimal fractions, for example $43 \div 5 = 8{\cdot}6$
Fractions, decimal fractions and percentages	I have investigated the everyday contexts in which simple fractions, percentages or decimal fractions are used and can carry out the necessary calculations to solve related problems. MNU 2-07a	• Uses knowledge of equivalent forms of common fractions, decimal fractions and percentages, for example, $\frac{3}{4} = 0{\cdot}75 = 75\%$, to solve problems. • Calculates simple percentages of a quantity, and uses this knowledge to solve problems in everyday contexts; for example, calculates the sale price of an item with a discount of 15%. • Calculates simple fractions of a quantity and uses this knowledge to solve problems; for example, find $\frac{3}{5}$ of 60.
Time	I can use and interpret electronic and paper-based timetables and schedules to plan events and activities, and make time calculations as part of my planning. MNU 2-10a	• Knows the relationships between commonly used units of time and carries out simple conversion calculations; for example, changes $1\frac{3}{4}$ hours into minutes. • Calculates durations of activities and events including situations bridging across several hours and parts of hours using both 12-hour clock and 24-hour notation.
Data and analysis	Having discussed the variety of ways and range of media used to present data, I can interpret and draw conclusions from the information displayed, recognising that the presentation may be misleading. MNU 2-20a	• Analyses, interprets and draws conclusions from a variety of data.

The activity tracker

1. See Resource 1.

   a Amman, Isla and Nuria buy new activity trackers in the sales. How much do they each pay?

   b What is the difference in price between the cheapest and most expensive sale prices?

2. Finlay decides he would like an activity tracker too (see Resource 2). How much did Finlay pay for his activity tracker?

3. The children set themselves the following daily steps targets (see Resource 3):

   By 5 pm their activity trackers display the following (see Resource 4):

   a How many more steps does Isla have to take to reach her target?

   b How many steps has Nuria completed so far?

   c How many steps has Amman completed so far?

d What is Amman's target for a fortnight?

e Nuria got up at 9 am this morning. What is her average **step count per hour** so far?

4. The bar chart shows the step count for one week (see Resource 5):

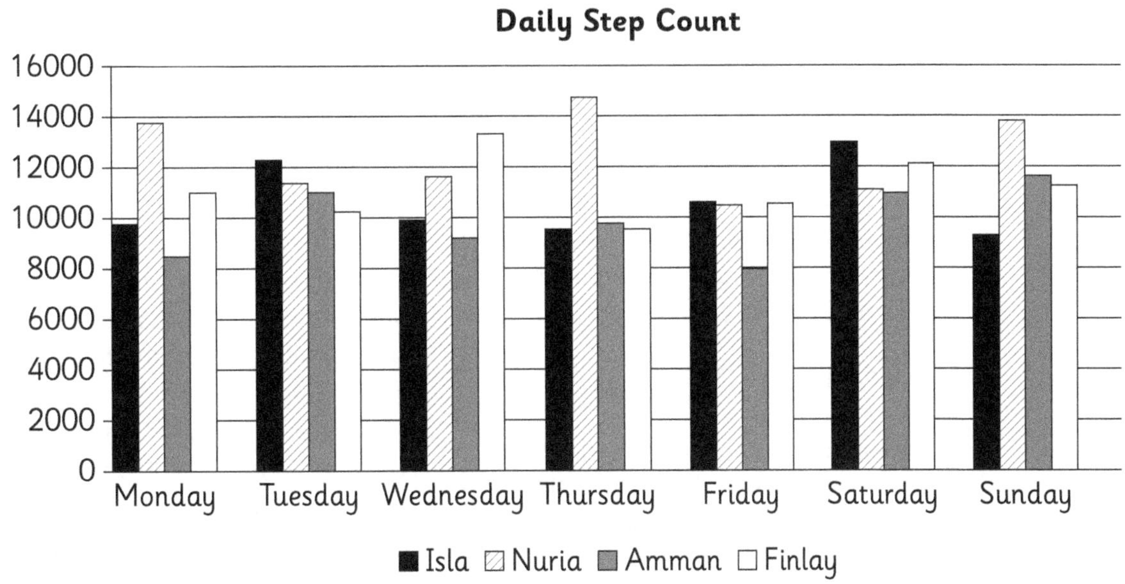

a On which days does Nuria beat her target?

b One of the children attends a running club every Tuesday, Saturday and Sunday. Which of the children attends the club? Justify your thoughts.

c Finlay beats his target four days out of the week. Estimate Finlay's target.

d Estimate, to the nearest thousand, the total weekly step count for each of the children. Order your estimates from most to least.

5. Isla got out of bed at 07:45 and climbed into bed at 20:10. For how long was Isla active today?

6. Nuria's activity tracker measures the quality of sleep she gets by recording Deep Sleep and REM Sleep. She falls asleep at exactly 8:15 p.m. When she wakes up, her activity tracker is showing the following data:

Deep Sleep: 6 hours 37 minutes

REM Sleep: 3¾ hours

At what time did Nuria wake up?

The trip

Curriculum organisers	Experiences and outcomes	Benchmarks
Estimation and rounding	I can use my knowledge of rounding to routinely estimate the answer to a problem, then, after calculating, decide if my answer is reasonable, sharing my solution with others. MNU 2-01a	• Applies knowledge of rounding to give an estimate to a calculation appropriate to the context.
Number and number processes	Having determined which calculations are needed, I can solve problems involving whole numbers using a range of methods, sharing my approaches and solutions with others. MNU 2-03a I have explored the contexts in which problems involving decimal fractions occur and can solve related problems using a variety of methods. MNU 2-03b	• Explains the link between a digit, its place and its value for numbers to three decimal places. • Adds and subtracts whole numbers and decimal fractions to two decimal places, within the number range 0 to 1 000 000. • Multiplies and divides whole numbers by multiples of 10, 100 and 1000. • Identifies familiar contexts in which negative numbers are used. • Orders numbers less than zero and locates them on a number line.
Money	I can manage money, compare costs from different retailers, and determine what I can afford to buy. MNU 2-09a	• Carries out money calculations involving the four operations.
Time	Using simple time periods, I can give a good estimate of how long a journey should take, based on my knowledge of the link between time, speed and distance. MNU 2-10c	• Reads and records time in both 12-hour and 24-hour notation and converts between the two. • Calculates durations of activities and events including situations bridging across several hours and parts of hours using both 12-hour clock and 24-hour notation. • Estimates the duration of a journey based on knowledge of the link between speed, distance and time.
Measurement	I can use the common units of measure, convert between related units of the metric system and carry out calculations when solving problems. MNU 2-11b	• Calculates the volume of cubes and cuboids in cubic centimetres (cm^3) and cubic metres (m^3). • Converts between common units of measurement using decimal notation; for example, 550 cm = 5·5 m; 3·009 kg = 3009 g.

End of second level assessment 2

The trip

The children are at the airport. They are going on a school trip to New York!

1.

✗ ✓ ✓ ✓ ✓

Baggage Allowance: 22 kg
Maximum dimensions (L × W × D)
90 cm × 60 cm × 30 cm

14·67 kg 7·38 kg 16·59 kg 5·24 kg

a Finlay and Amman are both taking two cases onto the plane. How much *over* or *under* the baggage allowance is each boy (see Resources 6 and 7)?

b What is the maximum allowable volume of an item of baggage (see Resource 6)? Give your answer in cubic centimetres.

9·45 kg ? ? 8·07 kg

c Isla's total baggage weight is under the allowance by 1 kg 720 grams (see Resource 8). How much does Isla's spotty bag weigh?

d Nuria's total baggage weight exceeds the allowance by 570 grams. How much does Nuria's black bag weigh?

2. The children go to the shop to buy supplies for the flight:

   a Finlay buys two bottles of water costing 79p each and three bars of chocolate costing 75p each. How much does he pay altogether?

   b Amman buys three cartons of apple juice and four packets of sweets. Each packet of sweets costs 90p and he pays £5·55 altogether. How much does one carton of apple juice cost?

3. The flight to New York is delayed by 1 hour 45 minutes (see Resource 9). What time will the flight depart?

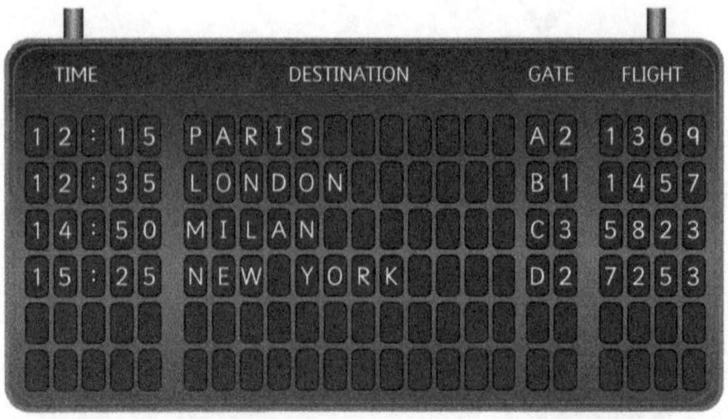

4. Finlay notices a display that shows the outside temperature during the flight is –54 °C.

   a The temperature was 17 °C when they left Glasgow. What is the difference between the two temperatures?

   b The display increases by 75 °C by the time they land in New York. What is the temperature in New York? You may use a number line to help you.

5. a See Resource 10. The total distance from Glasgow to New York is 5179 km. How far have they travelled so far?

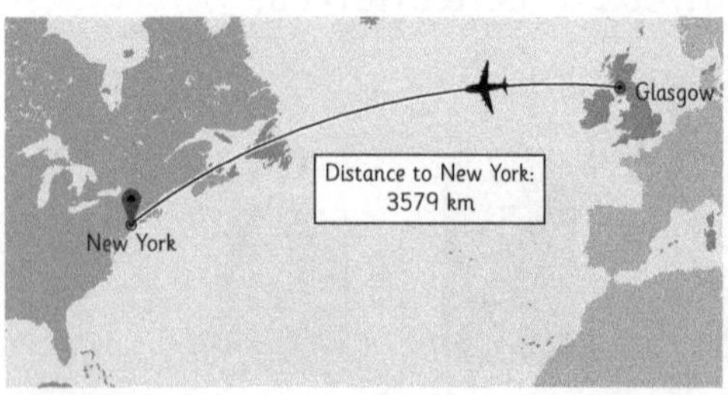

   b The plane travels at an average speed of 400 km per hour. Calculate how long they have travelled for so far.

   c Estimate how long the total journey will take.

   d The time in New York is 5 hours behind the time in Scotland. They arrive at 12:40 am (local time in New York). What time is it in Scotland?

   e How long did the flight take?

The escape room

Curriculum organisers	Experiences and outcomes	Benchmarks
Number and number processes	Having determined which calculations are needed, I can solve problems involving whole numbers using a range of methods sharing my approaches and solutions with others. MNU 2-03a	• Applies the correct order of operations in number calculations when solving multi-step problems. • Uses multiplication and division facts to the 10th multiplication table.
Multiples, factors and primes	Having explored the patterns and relationships in multiplication and division, I can investigate and identify the multiples and factors of numbers. MTH 2-05a	• Identifies multiples and factors of whole numbers and applies knowledge and understanding of these when solving relevant problems in number, money and measurement.
Measurement	I can use the common units of measure, convert between related units of the metric system and carry out calculations when solving problems. MNU 2-11b I can explain how different methods can be used to find the perimeter and area of a simple 2D shape or volume of a simple 3D object. MNU 2-11c	• Calculates the perimeter of simple straight-sided 2D shapes in millimetres (mm), centimetres (cm) and metres (m). • Calculates the area of squares, rectangles and right-angled triangles in square millimetres (mm^2), square centimetres (cm^2) and square metres (m^2). • Calculates the volume of cubes and cuboids in cubic centimetres (cm^3).
Patterns and relationships	Having explored more complex number sequences, including well-known named number patterns, I can explain the rule used to generate the sequence, and apply it to extend the pattern. MTH 2-13a	• Explains and uses a rule to extend well-known number sequences including square numbers, triangular numbers and Fibonacci sequence. • Applies knowledge of multiples, square numbers and triangular numbers to generate number patterns.
Expressions and equations	I can apply my knowledge of number facts to solve problems where an unknown value is represented by a symbol or letter. MTH 2-15a	• Solves simple algebraic equations with one variable; for example, $a - 30 = 40$ and $4b = 20$.
Properties of 2D shapes and 3D objects	Through practical activities, I can show my understanding of the relationship between 3D objects and their nets. MTH 2-16b	• Describes 3D objects and 2D shapes using specific vocabulary including regular, irregular, diagonal, radius, diameter and circumference. Applies this knowledge to demonstrate understanding of the relationship between 3D objects and their nets.

End of second level assessment 3

The escape room

The children are taking part in an Escape Room Challenge. To escape, they must solve these problems. Can you help them?

1. | ? | ? | ? | ? | ? | ? |

   Identify a 6-digit number based on the following clues:

   a No two digits are the same.

   b The hundred thousands digit is a factor of the thousands digit.

   c The thousands digit multiplied by the ones digit equals 36.

   d The tens of thousands digit is one third of the product of the hundreds and ones digits.

   e The thousands digit is an odd number with 3 factors.

   f The tens digit is a factor of 13.

2. | ? | ? |

   Identify the numbers on the back of these cards based on the following clues:

   - The difference between the two numbers is 26.
   - The sum of the two numbers is 114.

3. | ? (grey) | ? (white) |

   Identify the numbers on the back of these number cards based on the following clues:

   The white value is 7 times larger than the grey value.

   The sum of the 2 values is 64.

End of second level assessment 3

4. 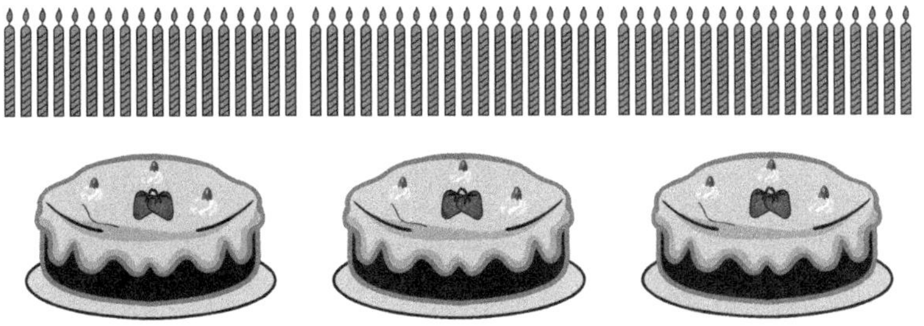

Finn, Eva and Rory all share the same birthday. Finn is three times older than Eva. Eva is half the age of Rory. Altogether they need 54 candles to decorate their birthday cakes (see Resource 11). How old is each child?

5. 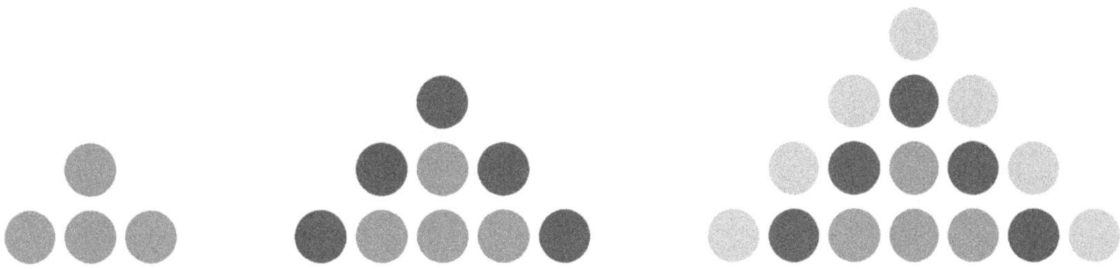

4 circles 9 circles 16 circles

The first three patterns of a sequence are shown (see Resource 12).

a Sketch the next pattern in the sequence.

b How many dots are in the 5th pattern?

c Write a rule for finding the number of dots.

6. 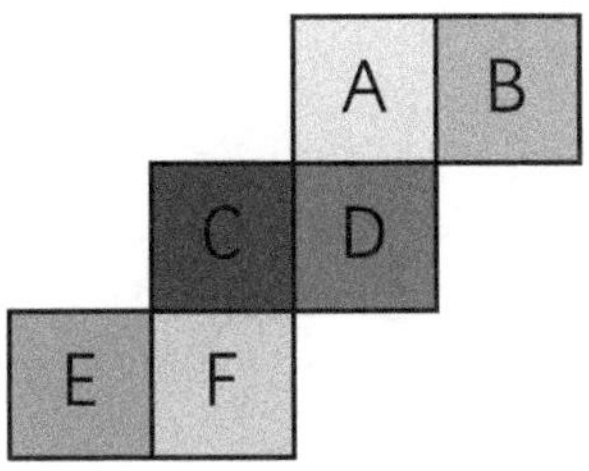

This net (see Resource 13) is folded to form a cube:

The cube is then positioned with face C on top.

i) Which face is on the bottom?

ii) How many different nets can you draw that form a cube?

7. Which of these 3D shapes can be placed in the tank without spilling any water (see Resource 14)? Justify your answer.

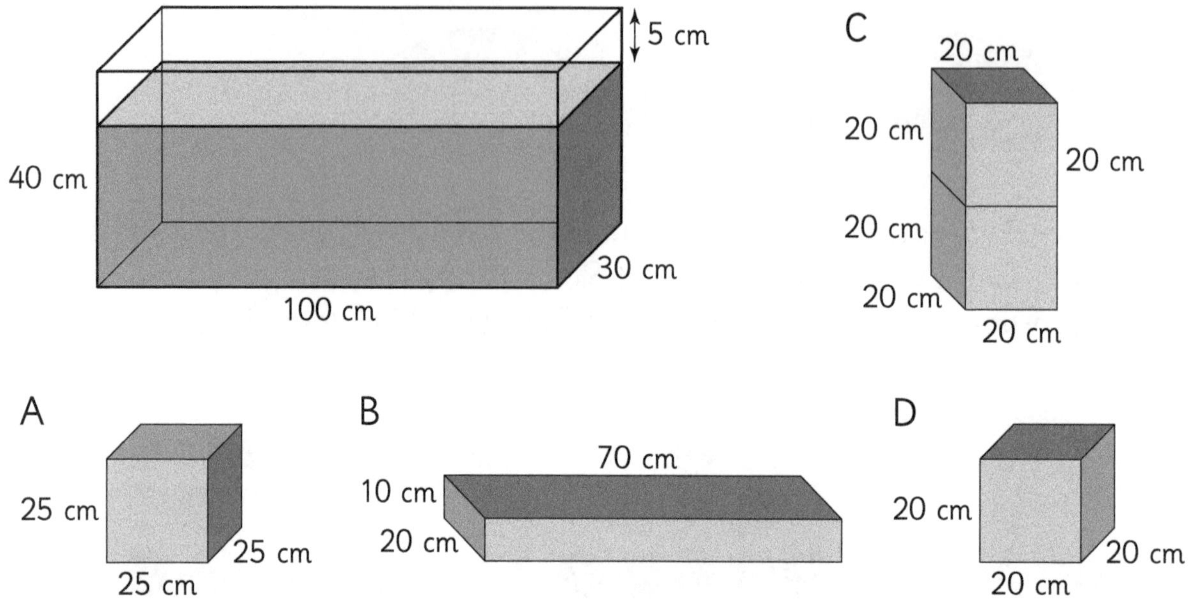

8. An irregular shape has been created using three different sizes of squares (see Resource 15). The sides of a medium square are double the length of the small square.

The sides of the large square are double the length of the medium square.

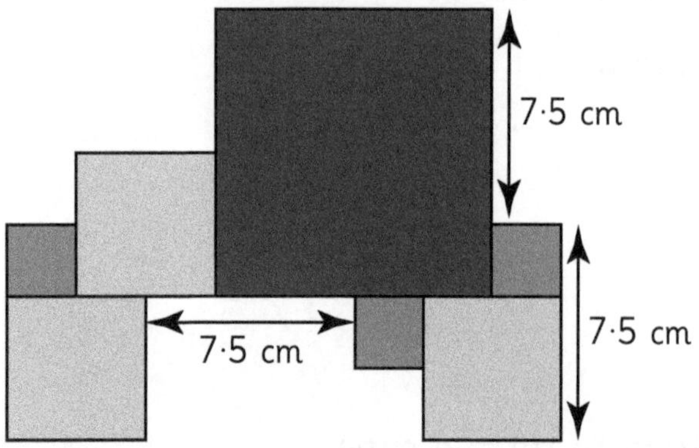

a What is the length of the sides of the small square?

b What is the total area of the shape?

c What is the perimeter of the whole shape?

d A total perimeter of 90 cm can be achieved by adding one further square. What size (small, medium or large) should be added and in which position?

End of second level assessment 4

The carnival

Curriculum organisers	Experiences and outcomes	Benchmarks
Fractions, decimal fractions and percentages	I have investigated the everyday contexts in which simple fractions, percentages or decimal fractions are used and can carry out the necessary calculations to solve related problems. MNU 2-07a	• Calculates simple fractions of a quantity and uses this knowledge to solve problems; for example, find $\frac{3}{5}$ of 60. • Expresses fractions in their simplest form.
Money	I can manage money, compare costs from different retailers, and determine what I can afford to buy. MNU 2-09a	• Carries out money calculations involving the four operations. • Compares costs and determines affordability within a given budget.
Time	I can use and interpret electronic and paper-based timetables and schedules to plan events and activities, and make time calculations as part of my planning. MNU 2-10a	• Uses and interprets a range of electronic and paper-based timetables and calendars to plan events or activities and solve real-life problems. • Calculates durations of activities and events including situations bridging across several hours and parts of hours using both 12-hour clock and 24-hour notation.
Angle, symmetry and transformation	I have investigated angles in the environment, and can discuss, describe and classify angles using appropriate mathematical vocabulary. MTH 2-17a	• Uses mathematical language including acute, obtuse, straight and reflex to describe and classify a range of angles identified within shapes in the environment. • Knows that complementary angles add up to 90° and supplementary angles add up to 180° and uses this knowledge to calculate missing angles. • Uses knowledge of the link between the eight compass points and angles to describe, follow and record directions. • Interprets maps, models or plans with simple scales; for example, 1 cm : 2 km.
Ideas of chance and uncertainty	I can conduct simple experiments involving chance and communicate my predictions and findings using the vocabulary of probability. MNU 2-22a	• Uses the language of probability accurately to describe the likelihood of simple events occurring; for example, equal chance; fifty-fifty; one in two; two in three; percentage chance; and $\frac{1}{6}$.

End of second level assessment 4

The carnival

1. The children are planning a trip from Haymarket to Glasgow to meet friends and visit the carnival (see Resource 16).

Edinburgh	10 d	1200	1215	1230	1245	1300	1315	1330	1345	1400	1415	1430	1445	1500	1515	1530	1545	1600	1615	1630	1645	1700	1715	1730
Haymarket	d	1206	1220	1235	1250	1305	1320	1334	1349	1404	1421	1436	1449	1505	1520	1534	1549	1604	1621	1634	1650	1705	1720	1734
Linlithgow	d	·	1233	·	1306	·	1333	·	1402	·	1433	·	1504	·	1533	·	1602	·	1634	·	1705	·	1734	·
Polmont	3 d	·	1238	·	1311	·	1337	·	1406	·	1438	·	1509	·	1538	·	1606	·	1638	·	1710	·	1739	·
Falkirk High	d	1223	1242	1253	1315	1325	1342	1352	1411	1422	1442	1453	1513	1522	1543	1552	1611	1622	1643	1652	1714	1723	1743	1752
Croy	3 d	1232	·	1302	·	1334	·	1401	·	1431	·	1502	·	1531	·	1601	·	1631	·	1701	·	1732	·	1801
Lenzie	3 a	*1247*	·	*1317*	·	*1347*	·	*1417*	·	*1449*	·	*1518*	·	*1547*	·	*1617*	·	*1647*	·	*1717*	·	*1747*	·	1806
Bishopbriggs	a	*1251*	·	*1321*	·	*1351*	·	*1421*	·	*1453*	·	*1522*	·	*1551*	·	*1621*	·	*1651*	·	*1722*	·	*1752*	·	*1823*
Glasgow Queen Street	10 a	1244	1302	1314	1333	1345	1403	1412	1430	1443	1502	1514	1531	1543	1605	1613	1635	1645	1703	1712	1733	1744	1806	1815

a What is the latest time the children can get the train from Haymarket if they want to arrive in Glasgow Queen Street before 3:00 pm?

b It is a 22 minute walk from Amman's house to the station. When should Amman leave his house if he wants to be at the station 5 minutes before the train leaves?

2. The children meet their friends at 3:00 pm at Glasgow Central:

Finlay points out that the hands on the clock form a right angle (see Resource 17).

a What size of angle does the minute hand turn through from 3:00 pm to 3:05 pm?

b What type of angle will the hands form at 3:30 pm?

c What size will this angle be?

3. They arrive at the carnival and find the following options for buying tickets:

- Child ticket: **£8·95**
- Adult ticket: **£12·49**
- Group ticket: **£9·85** per person
- Family ticket (2 adults & 2 children or 1 adult & 3 children): **£38·69**

There are 3 adults and 6 children altogether. What is the cheapest way to buy their tickets? Justify your choice.

4. The children want to go on the Waltzer first (see Resource 18).

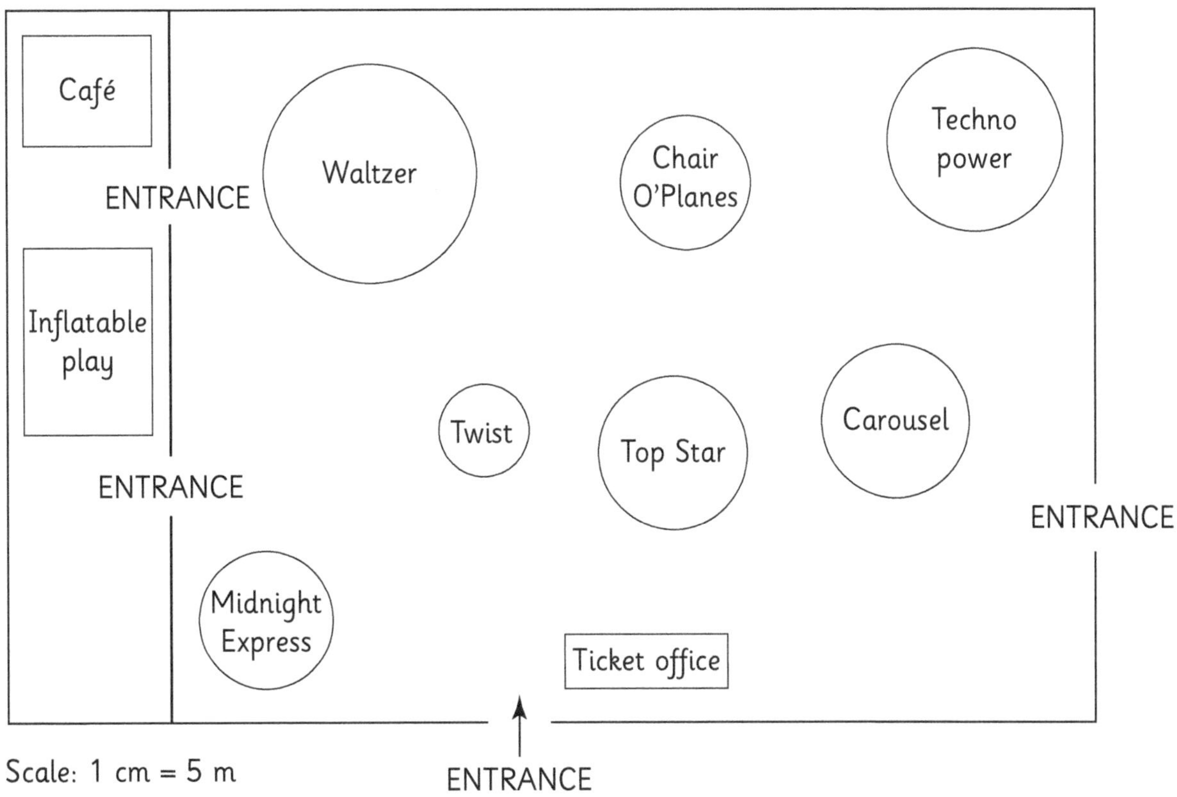

a Approximately how far do they have to walk from the ticket office?

b Using your knowledge of the compass points and/or angles, write directions to help the children get from the ticket office to the Waltzer.

5. The children have 20 coupons each that can be used on any ride or stall (2 coupons per attraction).

a Amman uses three fifths of his coupons on rides. How many rides does he go on?

b He gives the rest of his coupons to Isla and Nuria who share them fairly. How many coupons do each of the girls have now?

c Isla uses 16 coupons to go on rides. What fraction of her tickets does she have left for the stalls? State the fraction in its simplest form.

d Nuria uses three quarters of her coupons on rides. How many rides does she go on?

6. The children go to the food court. What do they buy (see Resource 19)?

a Finlay spends £4·71 on food and a drink.

b Amman bought food only. He handed over £5 and got 43p change.

c Nuria bought food and a drink. She handed over £10 and got £3·73 change.

d Isla spent three times as much on food as she did on a drink.

7. Finlay goes to a stall called Spin and Win. At the stall there is big wheel with black, grey and white segments (see Resource 20). If he guesses the correct colour as it is spun he wins a prize.

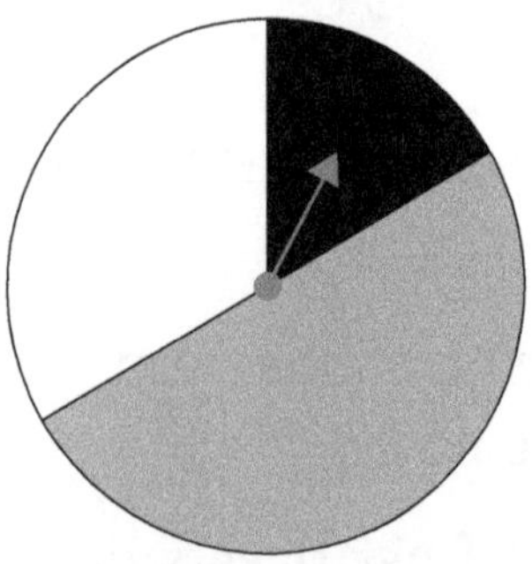

a What should he call to give him the best chance of winning a prize? Justify your answer by describing the chance of the spinner stopping on each of the colours.

b The results of the last ten people are shown:

Grey, White, Grey, Grey, White, Grey, Grey, White, Grey, Grey

What impact should this have on his decision? Justify your answer.

End of second level assessment 5

The festival

Curriculum organisers	Experiences and outcomes	Benchmarks
Money	I can use the terms profit and loss in buying and selling activities and can make simple calculations for this. MNU 2-09c	• Carries out money calculations involving the four operations. • Calculates profit and loss accurately; for example, when working with a budget for an enterprise activity.
Measurement	I can explain how different methods can be used to find the perimeter and area of a simple 2D shape or volume of a simple 3D object. MNU 2-11c I can use the common units of measure, convert between related units of the metric system and carry out calculations when solving problems. MNU 2-11b	• Converts between common units of measurement using decimal notation. • For example, 550 cm = 5·5 m; 3·009 kg = 3009 g. • Demonstrates understanding of the conservation of measurement; for example, draw three different rectangles each with an area of 24 cm^2. • Reads a variety of scales accurately.
Angle, symmetry and transformation	I have investigated angles in the environment, and can discuss, describe and classify angles using appropriate mathematical vocabulary. MTH 2-17a	• Interprets maps, models or plans with simple scales; for example, 1 cm : 2 km. • Uses knowledge of the link between the eight compass points and angles to describe, follow and record directions. • Describes, plots and records the location of a point, in the first quadrant, using coordinate notation. • Identifies and illustrates line symmetry on a wide range of 2D shapes and applies this understanding to complete a range of symmetrical patterns, with and without the use of digital technologies.

End of second level assessment 5

The festival

1. The children are visiting an island to attend a festival (see Resource 21).

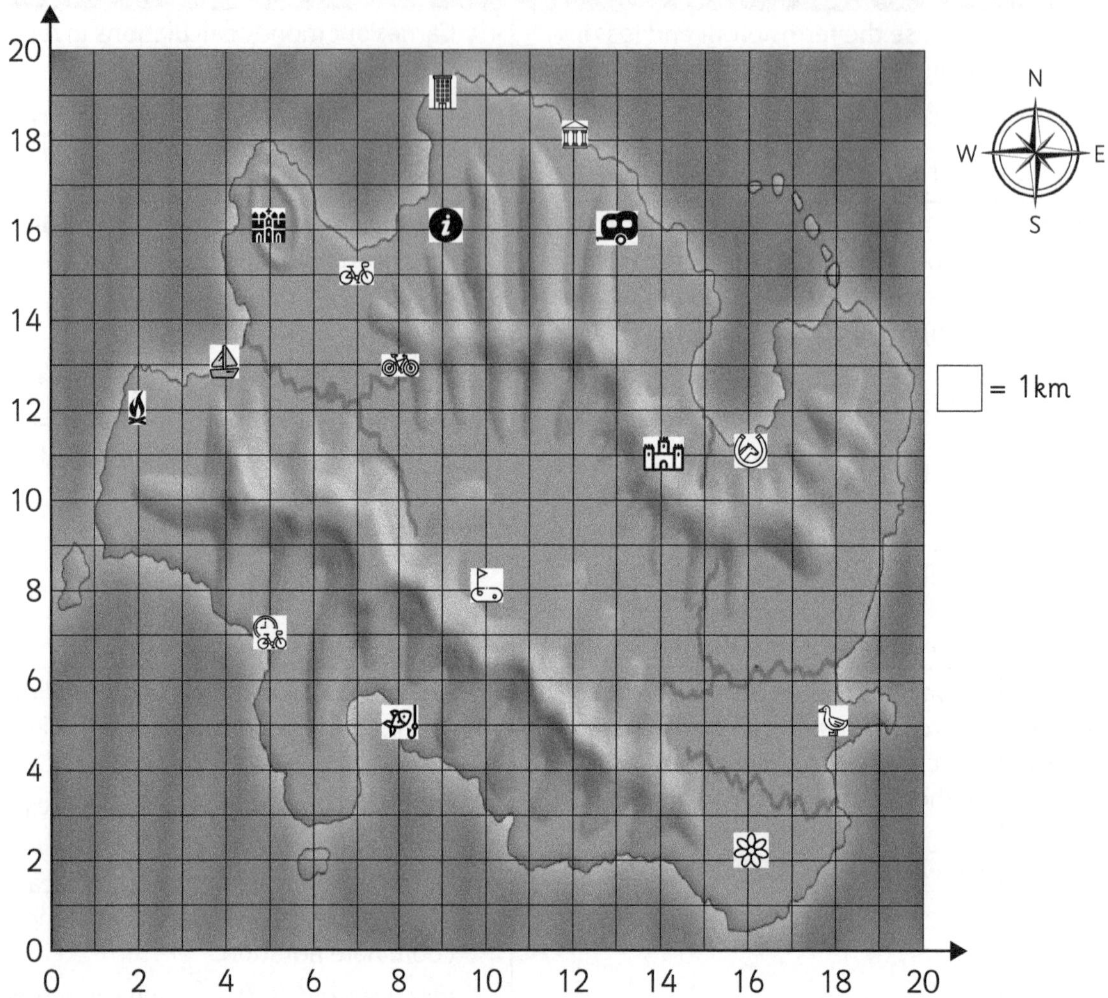

- Boat hire
- Building of historic interest
- Camp site
- Caravan site
- Castle or fort
- Cathedral or abbey
- Cycle hire
- Cycle trail
- Fishing
- Garden or arboretum
- Golf course or links
- Horse riding
- Information centre
- Mountain bike trail
- Museum
- Nature reserve

a The ferry sails from coordinate (2,1) towards the island on a bearing of north-east. What angle from north is the ferry travelling?

b Write the coordinates of the point where they will arrive on the island.

c They make their way to the fishing. From there they travel 2 km on a bearing of 000° then travel west for approximately 3 km. Where are they now?

d Give directions from the information centre to the castle (using your understanding of angles and / or compass points).

e The children notice that the picnic site is missing from the map. They are told that if they draw a straight line from the garden to the horse riding it forms a line of symmetry between the nature reserve and the picnic site. What are the coordinates of the picnic site?

2. The festival is taking place in the grounds of the castle, where the children are setting up a stall selling homemade lemonade. They are told to mark out a rectangular pitch with an area of 24 square metres using 1 metre lengths of fence:

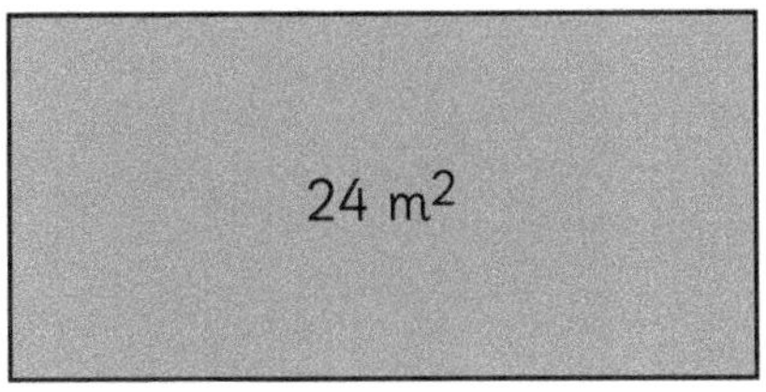

a What is the maximum perimeter of their pitch?

b What is the minimum perimeter of their pitch?

3. The children measure out the ingredients for their lemonade (see Resource 22).

a Write the mass in both grams and kilograms for the:

i Lemons

ii Caster sugar

b They fill the measuring jug with the amount of water shown ten times. How much water do they use?

End of second level assessment 5

4. They had to pay for the following:
   - 20 lemons – 50p each
   - Caster sugar – £2·79
   - Glass bottles – £28·50
   - Bags of ice – £9·99
   - Rental of pitch – £25

   They intend to sell the lemonade at £1·50 per cup. How many cups will they need to sell so that they do not make a loss? Show your calculations and justify your answer.
5. They sell a total of 70 cups of lemonade.

   a Do they make a profit or loss?

   b Calculate their profit or loss.

The activity day

Curriculum organisers	Experiences and outcomes	Benchmarks
Number and number processes	I have extended the range of whole numbers I can work with and, having explored how decimal fractions are constructed, can explain the link between a digit, its place and its value. MNU 2-02a Having determined which calculations are needed, I can solve problems involving whole numbers using a range of methods, sharing my approaches and solutions with others. MNU 2-03a I have explored the contexts in which problems involving decimal fractions occur and can solve related problems using a variety of methods. MNU 2-03b	• Reads, writes and orders whole numbers to 1 000 000, starting from any number in the sequence. • Reads, writes and orders sets of decimal fractions to three decimal places. • Applies the correct order of operations in number calculations when solving multi-step problems. • Adds and subtracts multiples of 10, 100 and 1000 to and from whole numbers and decimal fractions to two decimal places. • Multiplies and divides decimal fractions to two decimal places by 10, 100 and 1000.
Fractions, decimal fractions and percentages	I can show the equivalent forms of simple fractions, decimal fractions and percentages, and can choose my preferred form when solving a problem, explaining my choice of method. MNU 2-07b I have investigated how a set of equivalent fractions can be created, understanding the meaning of simplest form, and can apply my knowledge to compare and order the most commonly used fractions. MTH 2-07c	• Creates equivalent fractions and uses this knowledge to put a set of most commonly used fractions in order.

End of second level assessment 6

Curriculum organisers	Experiences and outcomes	Benchmarks
Time	I can use and interpret electronic and paper-based timetables and schedules to plan events and activities, and make time calculations as part of my planning. MNU 2-10a	• Knows the relationships between commonly used units of time and carries out simple conversion calculations; for example, changes 1½ hours into minutes.
Angle, symmetry and transformation	I can accurately measure and draw angles using appropriate equipment, applying my skills to problems in context. MTH 2-17b	• Knows that complementary angles add up to 90° and supplementary angles add up to 180° and uses this knowledge to calculate missing angles.

End of second level assessment 6

The activity day

The children have organised a day of activities: Lazer Tag, Go-karting and Virtual Golf.

At Lazer Tag, the children compete against other, either by themselves or together as a team. They compare their scores after the first game (see Resource 23):

1. Write the children's scores in order from first to last.
2. The scoring works as follows:
   - Direct hit **of** opponent: + 5000
   - Direct hit **of** opposing team captain: +10000
   - Direct hit **of** own team member: –250
   - Direct hit **by** opponent: –500
   - Direct hit **by** opposing team captain: –2000

During the second game the children score the following (see Resource 24):

	Finlay	Isla	Nuria	Amman
hits of opponent (+5000)	10	12	15	9
hits of opposing captain (+10000)	6	3	4	2
hits of team member (–250)	0	1	2	3
hits by opponent (–500)	8	5	7	6
hits by opposing captain (–2000)	1	2	0	3

a Calculate the boys' individual scores for the second game.

b Calculate the children's overall team score for the second round.

3. At the Go-karting, the children's average lap times are calculated for the first race as follows (see Resource 25):

Finlay	Isla	Nuria	Amman	Eva	Max	Petra	Ali
35·28 s	36·2 s	35·082 s	36 s	36·021 s	36·12 s	35·8 s	35·05 s

a Write the children's average lap times in order from fastest to slowest.

b What is the time difference between Isla and Finlay's average lap times?

c What would Finlay's average time be if the race was run for:

i. 10 laps

ii. 100 laps

iii. 1000 laps

4. The total times for the second race (10 laps) are as follows (see Resource 26):

Finlay	Isla	Nuria	Amman	Eva	Max	Petra	Ali
346·1 s	339·08 s	360s	343·25 s	337·7 s	365·1 s	358·68 s	341·65 s

a Convert Nuria's time into minutes.

b What are the children's average lap times?

c Finlay receives a 10% time penalty because of a false start. What is his new time?

5. The final race is a 30 second sprint to see how much of the track they can complete in that time. Here are the results:

Finlay: three quarters of the track.

Nuria: two thirds of the track.

Amman: five eighths of the track.

Isla: half of the track.

a Who is the fastest boy? Justify your answer.

b Who is the fastest girl? Justify your answer.

c Who is the fastest overall? Justify your answer.

6. At the Virtual Golf, the children make the following shots (see Resource 27):

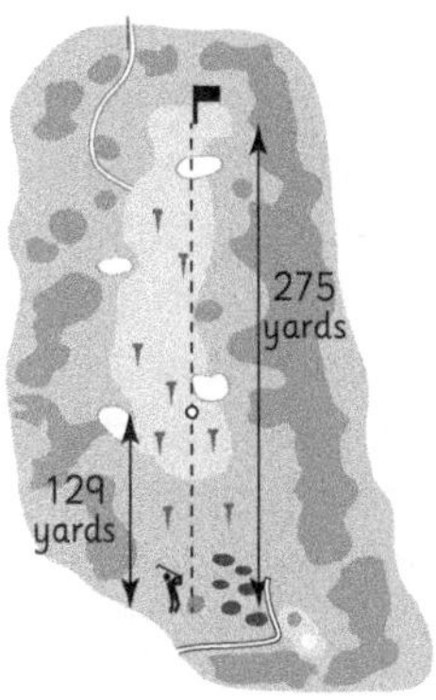

Finlay's 1st shot

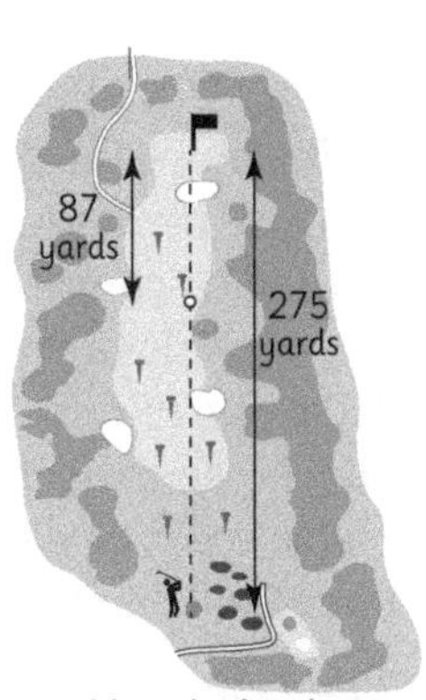

Nuria's 1st shot

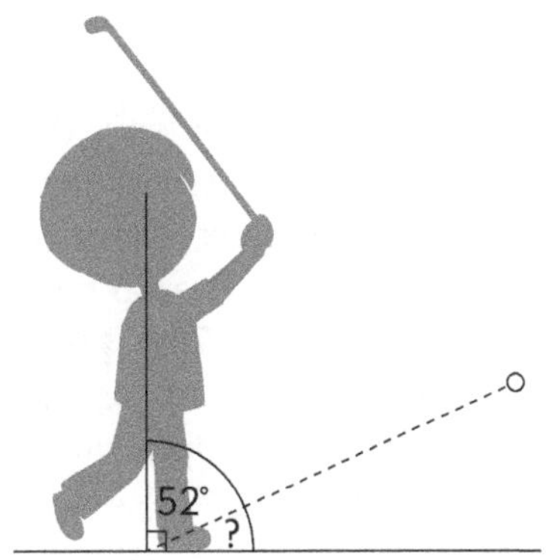

Amman's 1st shot

a How far is Finlay from the hole?

b What was the distance of Nuria's first shot?

c At what angle from the ground does Amman hit his shot? (Calculate the missing angle)

End of second level assessments: answers and marking guidance

Assessment 1: The activity tracker

<table>
<tr><th>Question</th><th>Topic</th><th>Question</th></tr>
<tr>
<td>Q1a</td>
<td>Fractions, decimal fractions and percentages

MNU 2-07a</td>
<td>

Answer
Isla: £38.75; Nuria: £41.25; Amman £37.80

On track
There is evidence to show that the child can solve a fraction / percentage of a value problem. The relationships between the values in each problem can be analysed as follows:

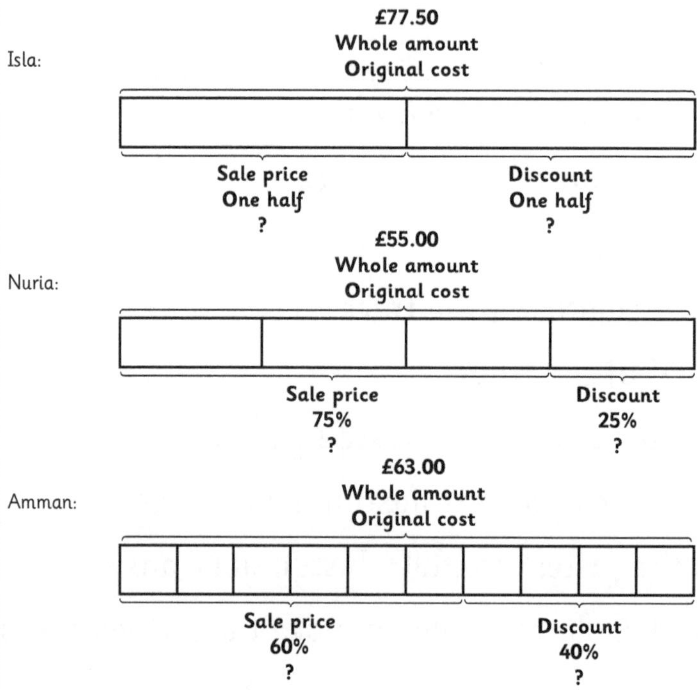

There may be evidence that the child understands that fraction / percentage problems are interchangeable. For example, it is not necessary to find Nuria's sale price using percentage only. This could be converted into a fraction and solved appropriately. The bar model above provides a visual representation of this as we can see that 75% is represented by three out of the four equal parts, i.e. three quarters. Based on this analysis, pupils can then choose an effective strategy for solving, for example, Nuria's sale price could be calculated by sharing £55 between the four parts (£55 ÷ 4 = £13.75) and then finding three lots of £13.75:

- £13.75 + £13.75 + £13.75 = ? OR
- 3 × £13.75 = ?

Review
Provide pupils with opportunities to analyse and solve fraction, decimal and percentage problems using a range of contexts in order that they understand the inherent link between them and that they are interchangeable, i.e. 'I can solve a percentage problem using the equivalent fraction and vice versa'.

</td>
</tr>
</table>

End of second level assessments: answers and marking guidance

<table>
<tr><th>Question</th><th>Topic</th><th>Question</th></tr>
<tr><td>Q1b</td><td>Number & number processes

MNU 2-03a</td><td>

Answer
£3.45

On track
There is evidence that pupils understand the relationship between the values in the question, i.e. We are comparing the two costs:

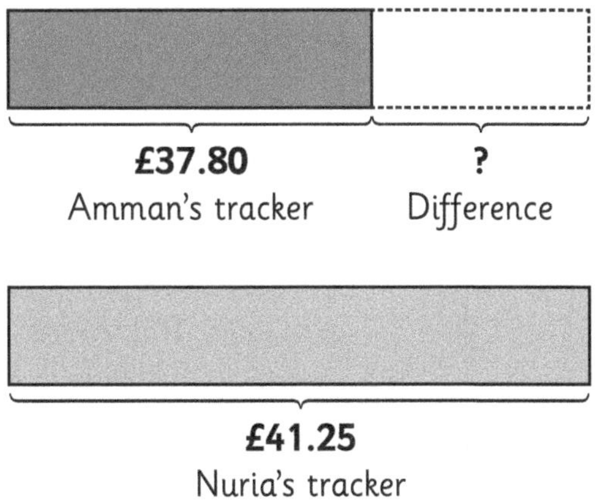

This representation allows us to see that the difference can be found by:

- subtracting £37.80 from the cost of Nuria's tracker (£41.25 – £37.80 = ?); OR
- counting back from £41.25 to the cost of Amman's tracker (£41.25 – ? = £37.80); OR
- counting on from £37.80 to the cost of Nuria's tracker (£37.80 + ? = £41.25). For example, they may choose to count on as illustrated:

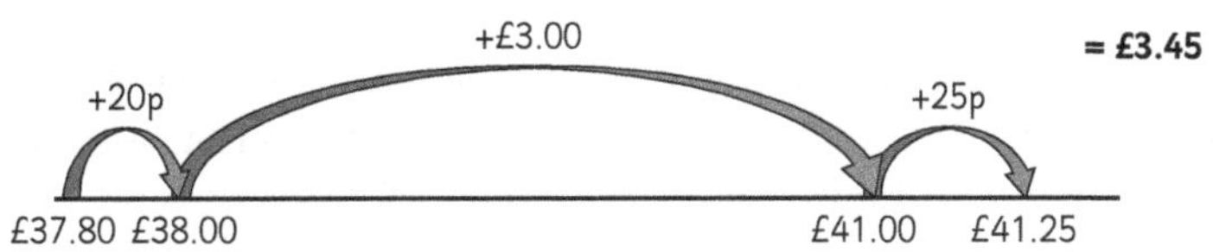

Review
Provide pupils with opportunities to analyse and solve comparison problems involving decimal numbers where the unknown value is either the 'difference' or one of the 'compared values'. Use a range of contexts; for example, height, length, money.

</td></tr>
<tr><td>Q2</td><td>Fractions, decimal fractions and percentages

MNU 2-07a</td><td>

Answer
£45

On track
There is evidence that the child can analyse the problem and apply a suitable strategy for finding the sale price. For example, the problem can be analysed as follows:

</td></tr>
</table>

End of second level assessments: answers and marking guidance

<table>
<tr><th>Question</th><th>Topic</th><th>Question</th></tr>
<tr><td></td><td></td><td>
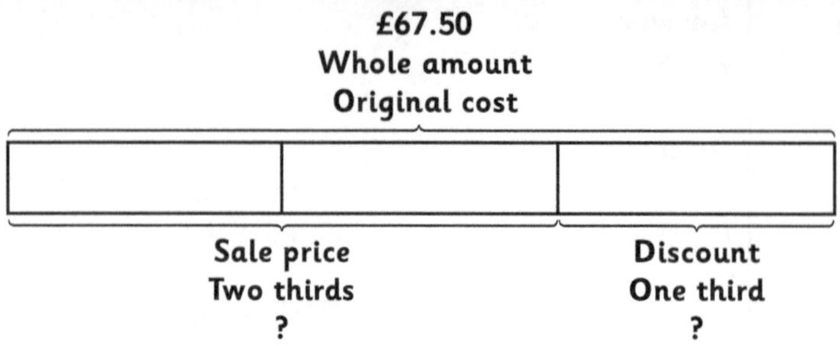

This analysis allows us to see that the sale price can be found by:

<ul>
<li>Dividing the original cost by 3; this may be achieved by partitioning, for example, by sharing £60 between each part (£20), then sharing £6 between each part (£2), then sharing the remaining £1.50 between each part (50p). This results in each third being worth £22.50.</li>
<li>Doubling the value of one third to find the value of two thirds i.e. the sale price (2 x £22.50 = £45)</li>
</ul>

Review
Explore the link between fractions and the whole so that the child understands that by saving one third of the price, Finlay is paying two thirds. Provide pupils with opportunities to analyse and solve fraction, decimal and percentage problems using a range of contexts in order that they understand the inherent link between them and that they are interchangeable, i.e. 'I can solve a percentage problem as a fraction and vice versa'.
</td></tr>
<tr><td>Q3a</td><td>Number & number processes

MNU 2-03a</td><td>
Answer: 2418 steps

On track
There is evidence that the child can analyse the problem and apply a suitable strategy for finding the number of steps Isla needs to meet her target, for example, the problem can be analysed as follows:

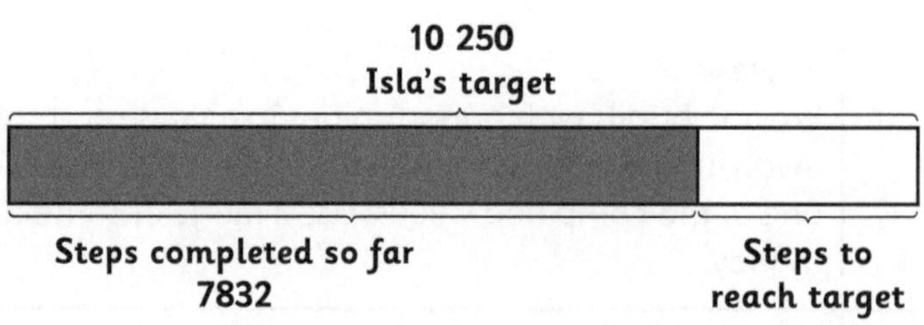

This representation allows us to see that the unknown part can be found by:

<ul>
<li>subtracting 7832 steps from Isla's target (10250 – 7832 = ?); OR</li>
<li>counting back from 10250 steps to the steps completed so far (10250 – ? = 7832); OR</li>
</ul>
</td></tr>
</table>

Question	Topic	Question
		• counting on from 7832 steps to Isla's target (7832 + ? = 10250). For example, they may choose to count on as illustrated: 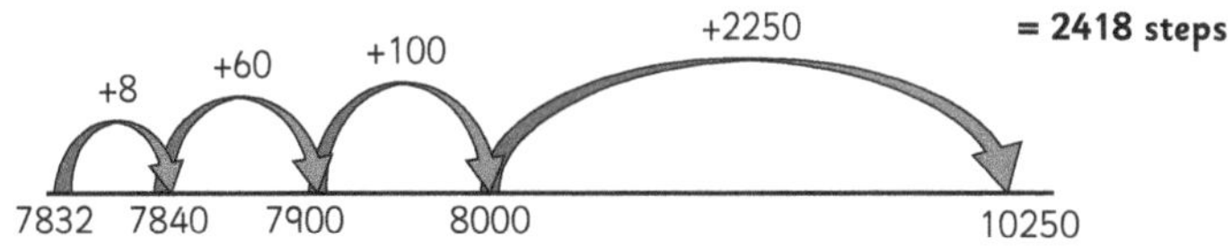 **Review** Provide pupils with opportunities to analyse and solve part-part-whole problems where the unknown value is either the 'whole' or one of the 'parts', using a range of contexts.
Q3b	Fractions, decimal fractions and percentages MNU 2-07a	**Answer**: 9600 steps **On track**: Identifies that they need to calculate 80% of the target and can choose an appropriate strategy. For example, the solution can be found by finding 8 tenths of 12 000, 4 fifths of 12 000 or by finding 2 tenths and subtracting from 12 000. Analysing the problem effectively using a bar model would demonstrate the necessary understanding: 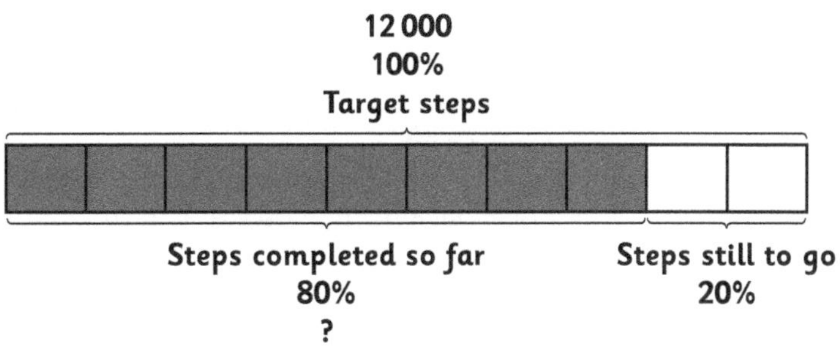 From here, the child should choose a suitable strategy for dividing 12 000 between each 10% block (for example, using partitioning) and multiplying by the 8 parts relating to the 'steps completed so far'. Or by multiplying 12 000 by 80 'over' 100. **Review** Revise finding percentages or fractions, perhaps looking at 10% and $\frac{1}{10}$ 'blocks' combined with visual models as shown above.

<table>
<tr><th>Question</th><th>Topic</th><th>Question</th></tr>
<tr><td>Q3c</td><td>Number & number processes

MNU 2-03a</td><td>

Answer: 7626 steps

On track
There is evidence that the child can analyse the problem and apply a suitable strategy for finding the number of steps Amman has completed so far. For example:
The problem can be analysed as follows:

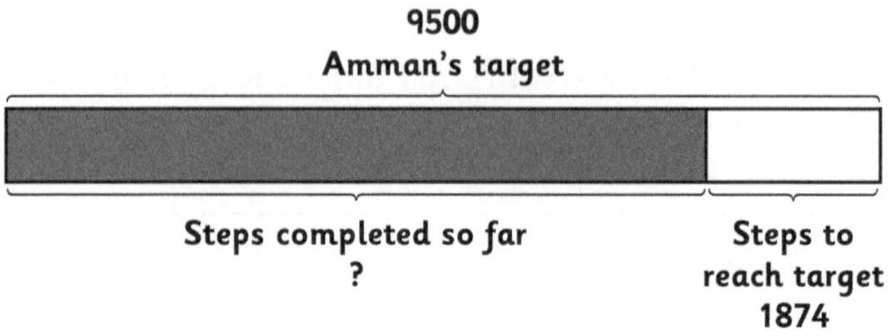

This representation allows us to see that the unknown part can be found by:

- subtracting 1874 steps from Amman's target (9500 – 1874 = ?): OR

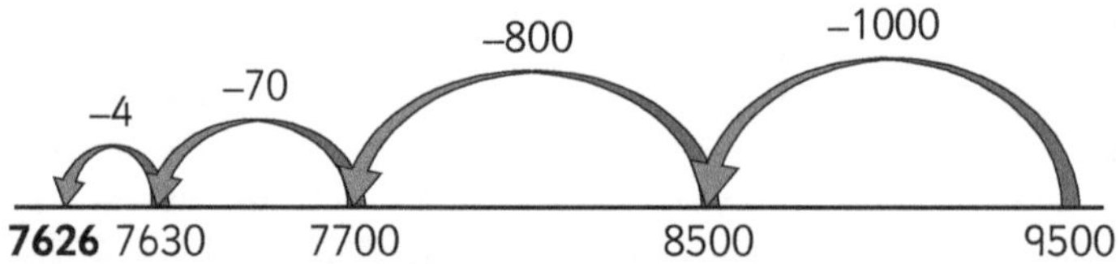

- counting back from 9500 steps to the 'steps to reach target' (9500 – ? = 1874): OR

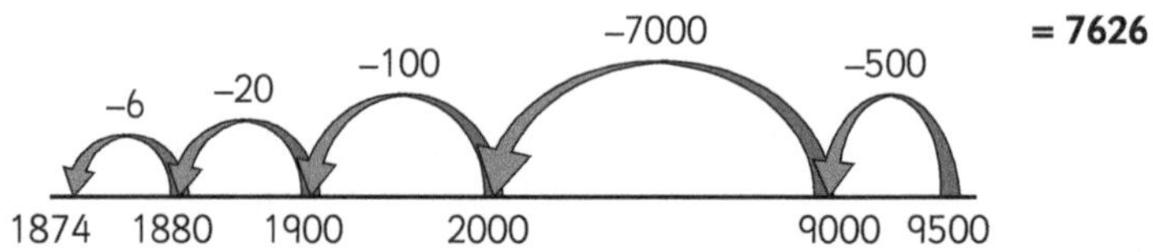

- counting on from 1874 steps to Amman's target (1874 + ? = 9500):

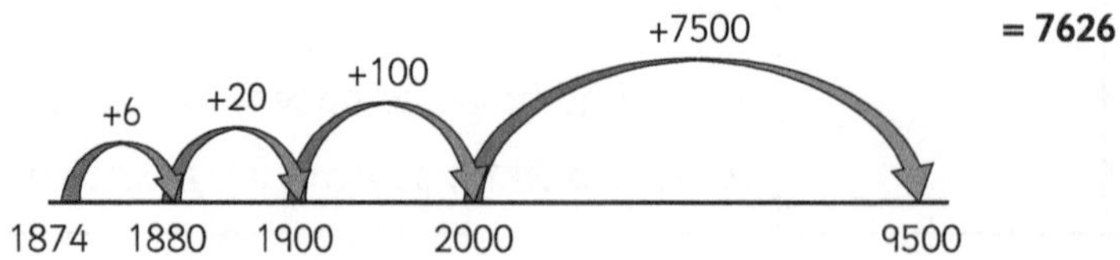

Review
Provide pupils with opportunities to analyse and solve part-part-whole problems where the unknown value is either the 'whole' or one of the 'parts', using a range of contexts.

</td></tr>
</table>

End of second level assessments: answers and marking guidance

Question	Topic	Question
Q3d	Number & number processes MNU 2-03a	**Answer** 133 000 steps **On track** There is evidence that the child can identify and apply a suitable strategy for multiplying Amman's daily target by 14 days; for example, by drawing an array: 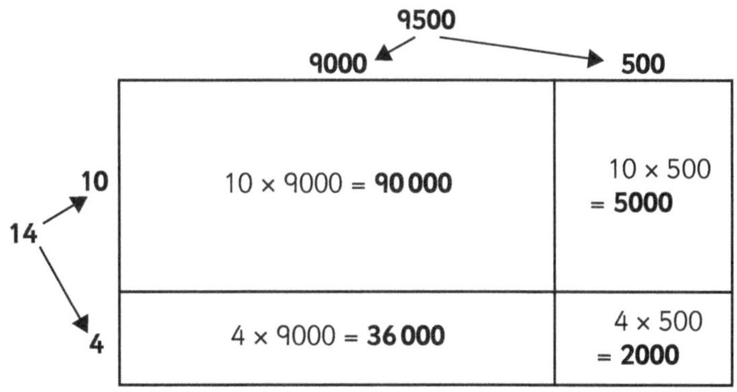 90 000 + 36 000 + 5000 + 2000 = **133 000** **Review** Explore strategies for solving with the child. For example, they may multiply 9500 steps by 7 days and then double. Check that pupils have the necessary understanding of place value to be able to partition and then multiply by multiples of 10, 100, 1000, etc.
Q3e	Number & number processes MNU 2-03a	**Answer** 1200 steps per hour **On track** There is evidence that the child can determine the number of hours that Nuria has been active (using a suitable strategy) and understands that the hourly average can be found by dividing her 'step count so far' (9600 steps) by the number of hours she has been active (8 hours). **Review** Ensure child is able to calculate duration using appropriate strategies (counting on, counting back, etc), understands the concept of 'average' and can identify and apply a suitable division strategy.
Q4a	Data and analysis MNU 2-20a	**Answer** Monday, Thursday and Sunday **On track** There is evidence that the child can interpret the information on the bar chart, i.e. that the stripy bars relate to Nuria; that they can read the scale on the Y axis; and that they can identify the bars that are over 12 000 steps. **Review** Ensure that the child can read and interpret information from charts and graphs and can construct a bar chart given a set of information.

End of second level assessments: answers and marking guidance

Question	Topic	Question
Q4b	Data and analysis MNU 2-20a	**Answer** Amman because on these days he was well over his target. **On track** There is evidence that the child can identify and interpret the relevant information, i.e. they can identify each child's results for the days in question and understand that there will be 'spikes' for a specific child on those days. **Review** Ensure that the child can interpret information from a bar chart and justify their responses.
Q4c	Data and analysis MNU 2-20a	**Answer** Finlay's target is 11 000 (accept anything between 10 400 and 11 400). **On track** There is evidence that the child is able to identify Finlay's top 4 days and provide an estimate somewhere in the range 10 400 and 10 500. **Review** Ensure that the child can interpret data from a bar chart and is able to estimate / round to the nearest 1000.
Q4d	Data and analysis MNU 2-20a	**Answer** Accept estimates within the following ranges – Isla: 72 000 to 82 000; Nuria: 85 000 to 95 000; Amman: 68 000 to 78 000; Finlay: 76 000 to 86 000 **On track** There is evidence that the child can interpret the relevant information from the bar chart, round each bar to the nearest 1000 and add up the resultant estimates for each of the four children. **Review** Ensure that the child can interpret data from a bar chart estimate / round to the nearest 1000 and add estimates using an appropriate strategy.
Q5	Data and analysis MNU 2-20a	**Answer** 12 hours and 25 minutes **On track** There is evidence that the child understands time displayed in 24-hour format and can calculate duration using an appropriate strategy (counting on, counting back, etc). **Review** Ensure child can demonstrate their thinking; for example, they may use a number line to count on from 7:45 by counting on 15 minutes, then 12 hours, then a further 10 minutes.

Question	Topic	Question
Q6	Time MNU 2-10a	**Answer** 6:37 a.m. **On track** There is evidence that the child can convert times given in fractions into hours and minutes (i.e. ¾ of an hour is equal to 45 minutes) and can apply this understanding to solve duration problems appropriately. **Review** Ascertain whether the child can find an efficient strategy; for example, it will be easier to start at 8:15 and count on the 3 ¾ hours to get to 12 midnight before simply counting on 6 hours 37 minutes. Ensure that the child can convert ¼ hour into 15 minutes, ¾ hour into 45 minutes, etc.

End of second level assessments: answers and marking guidance

Assessment 2: The trip

Question	Topic	Question
Q1a	Measurement MNU 2-11b	**Answer** Finlay: 0.05 kg (or 50 g) over; Amman: 0.17 kg (or 170 g) under **On track** There is evidence that the pupil can identify, and correctly use, a suitable strategy to solve the problem. For example, they may find the total mass of Finlay's bags using a partitioning strategy: 14 kg + 7 kg = 21 kg 0.6 kg + 0.3 kg = 0.9 kg 0.07 kg + 0.08 kg = 0.15 kg 21 kg + 0.9 kg + 0.15 kg = 22.05 kg OR by converting kilograms into grams, then partitioning: 14 kg + 7 kg = 21 kg 600 g + 300 g = 900 g 70 g + 80 g = 150 g 21 kg + 900 g + 150 g = 22 kg 50 g Once the have found the total weight of Finlay's baggage they can then find the difference between this and the total baggage allowance by: Subtracting the baggage allowance from the total combined mass: 22.05 kg – 22 kg = ? Counting on: 22 kg + ? = 22.05 kg OR Counting back: 22.05 kg – ? = 22 kg **Review** Check the pupil understands the relationship between grams and kilograms and can interpret, for example, 14.67 kg as 14 whole kilograms and 670 *thousandths* of a kilogram. Check the pupil understands how to convert kilograms to grams, i.e. by dividing by 1000. Provide opportunities for pupils to solve addition and subtraction problems that require them to connect decimal fractions written to two decimal places with thousandths; for example, within the contexts of mass and capacity.
Q1b	Measurement MNU 2-11b	**Answer** 162 000 cm^3 **On track** There is evidence that pupils understand the term cubic centimetre and appreciate that they need to multiply the dimensions of the bag (length × width × depth) to calculate its volume, i.e. 90 cm × 60 cm × 30 cm. **Review** Check the pupil understand that the volume of space and solids is measured in cubes and that a Dienes 'one' has a volume of 1 cubic centimetre. Provide opportunities for multiplying by multiples of ten using concrete materials and visual representations.

End of second level assessments: answers and marking guidance

Question	Topic	Question
Q1c	Measurement MNU 2-11b	**Answer** 10.83 kg (or 10 kg 830 g) **On track** There is evidence that pupils understand the relationship between the values in the question, i.e. 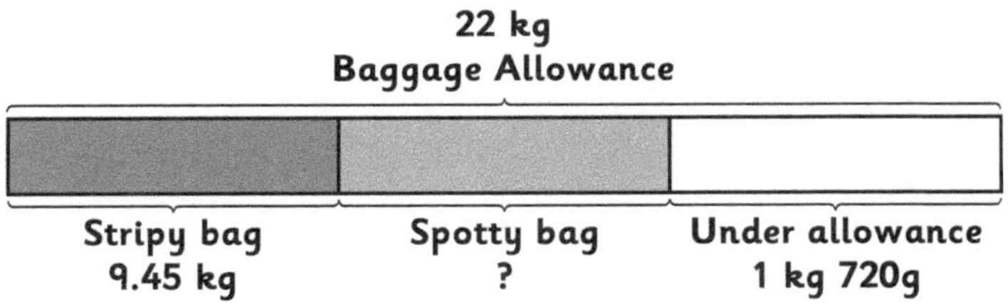 This representation allows us to see that the unknown part (the spotty bag) can be found by: • subtracting 9.45 kg and 1 kg 720 g (1.72 kg) from 22 kg in two steps, i.e. by subtracting one known value from 22kg and the other known value from the result; OR • adding the two parts we know (9.45 kg and 1kg 720 g) and finding the difference between this and the whole amount (the baggage allowance) For example, the pupil may choose to add 9.45 kg and 1.72 kg to get 11.17 kg. From here they may decide to count on to 22 kg: 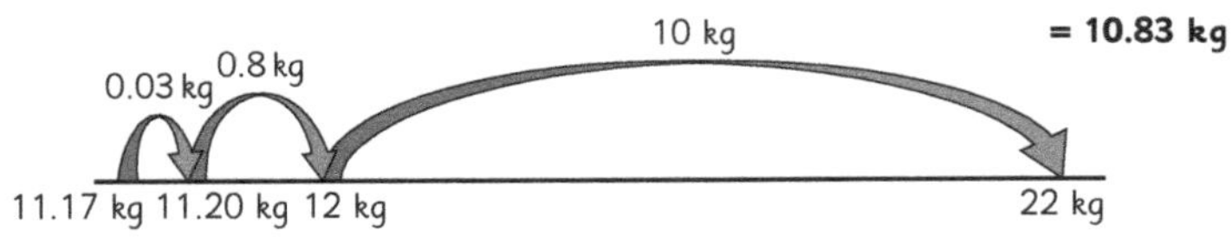 **Review** Check the pupil understands how to convert kilograms to grams and vice-versa. Provide them with opportunities to analyse and solve multi-step problems within the context of mass, where the unknown value is either the 'whole' or one of the 'parts'.
Q1d	Measurement MNU 2-11b	**Answer** 14.5 kg (or 14 kg 500 g) **On track** There is evidence that pupils understand the relationship between the values in the question. i.e. 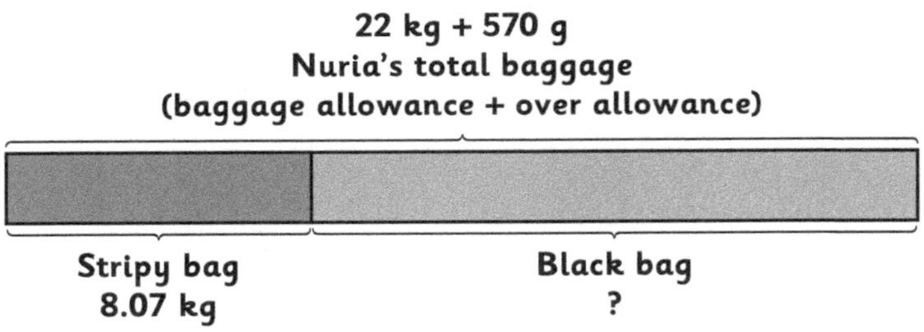

End of second level assessments: answers and marking guidance

<table>
<tr><th>Question</th><th>Topic</th><th>Question</th></tr>
<tr><td></td><td></td><td>
This representation allows us to see that the unknown part (the black bag) can be found by:

- subtracting 8.07 kg from the sum of the baggage allowance and the amount by which Nuria has exceeded the allowance (22 kg + 570 g); OR
- counting back from 22 kg 570 g (22.57 kg) to 8 kg 70 g (8.07 kg); OR
- counting on from 8.07 kg to the sum of the baggage allowance and the amount by which Nuria has exceeded the allowance (22 kg + 570 g); for example, they may choose to count on as illustrated:

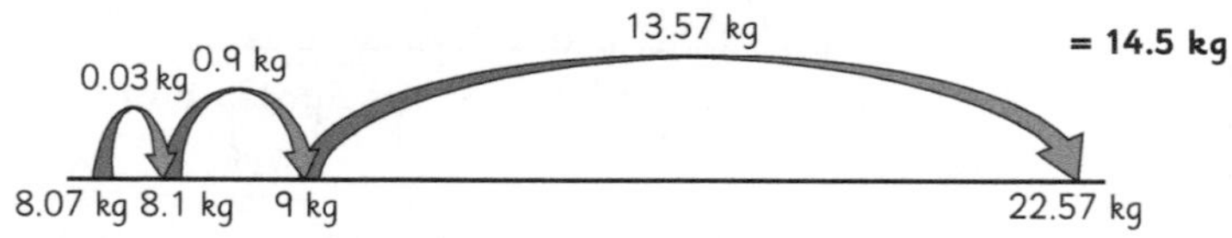

Review
Provide pupils with opportunities to analyse and solve multi-step problems within the context of mass, where the unknown value is either the 'whole' or one of the 'parts'. If necessary, provide further experiences in converting between the appropriate units of measurement, i.e. kilograms and grams.
</td></tr>
<tr><td>Q2a</td><td>Money

MNU 2-09a</td><td>
Answer
£3.83

On track
There is evidence that pupils understand the relationship between the values in the question. i.e.

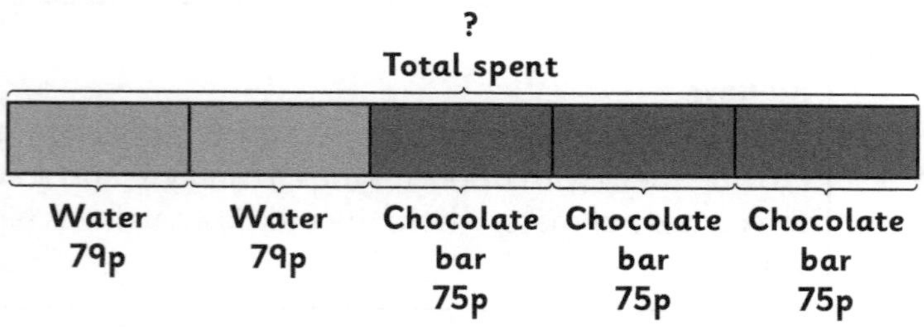

This representation allows us to see that the total spent can be found by adding the cost of the five items.

There are a range of strategies that may be used to solve this. For example, it may be more efficient to:

1. round and adjust to find the cost of the water, i.e. double 80p (to get £1.60) then subtract 2p to get £1.58
2. multiply to find the cost of the chocolate, i.e. 3 × 75p = (3 × 70) + (3 × 5) = 210p + 15p = 225p = £2.25 OR 75p × 2 = £1.50; £1.50 + 75p = £2.25
3. add £2.25 + £1.58 by partitioning mentally, i.e. (£2 + £1) + (25p + 58p) = £3 + 83p = £3.83.
</td></tr>
</table>

<table>
<tr><th>Question</th><th>Topic</th><th>Question</th></tr>
<tr><td></td><td></td><td>

Review
Provide the pupil with opportunities to analyse and solve multi-step problems where the unknown value is either the 'whole' or one of the 'parts'. Use contexts that require them to work with numbers to two decimal places; for example, height and money. If necessary, provide further experiences of converting between the appropriate units, i.e. pounds and pence; metres and centimetres.

</td></tr>
<tr><td>Q2b</td><td>Money

MNU 2-09a</td><td>

Answer
65p

On track
There is evidence that pupils understand the relationship between the values in the question. i.e.

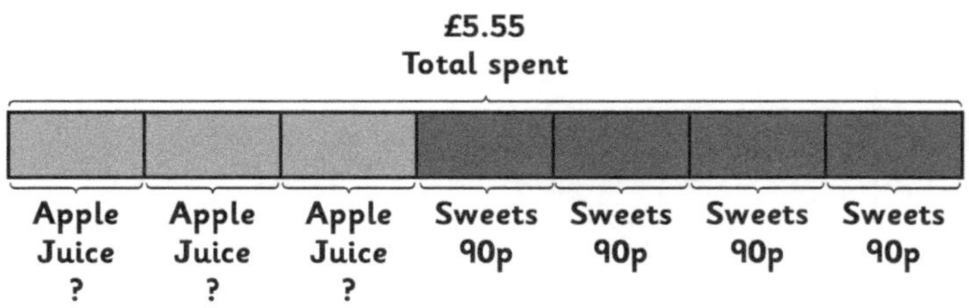

There are a range of strategies that may be used to solve this. For example, the pupil may decide to:

1. find the total costs of the sweets (4 x 90p = £3.60);
2. count on from £3.60 to £5.55 to find the total cost of the three apple juices (£1.95);
3. divide £1.95 by 3 to find the cost of one apple juice.

Review
Provide the pupil with opportunities to analyse and solve multi-step problems where the unknown value is either the 'whole' or one of the 'parts'. Use contexts that require them to work with numbers to two decimal places; for example, height and money. If necessary, provide further experiences of converting between the appropriate units, i.e. pounds and pence; metres and centimetres.

</td></tr>
<tr><td>Q3</td><td>Time

MNU 2-10c</td><td>

Answer
17:10 (5:10 p.m.)

On track
There is evidence that pupils understand that the flight will leave 1 hour and 45 minutes after the intended departure time. The pupil chooses a suitable strategy for calculating the new departure time; for example, by counting on 1 hour 45 minutes from 15:25.

</td></tr>
</table>

End of second level assessments: answers and marking guidance

Question	Topic	Question
		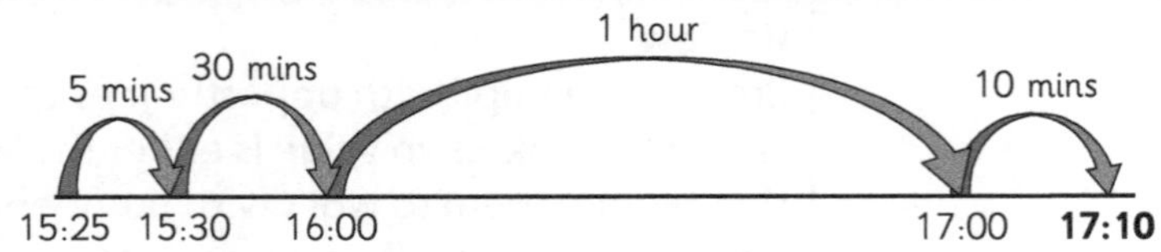 **Review** Provide opportunities to solve time duration problems using both 12-hour and 24-hour times.
Q4a	Number & number processes MNU 2-03a	**Answer** 71 °C **On track** There is evidence that pupils are able to find the difference between a positive and negative number, for example, by counting on using a number line: 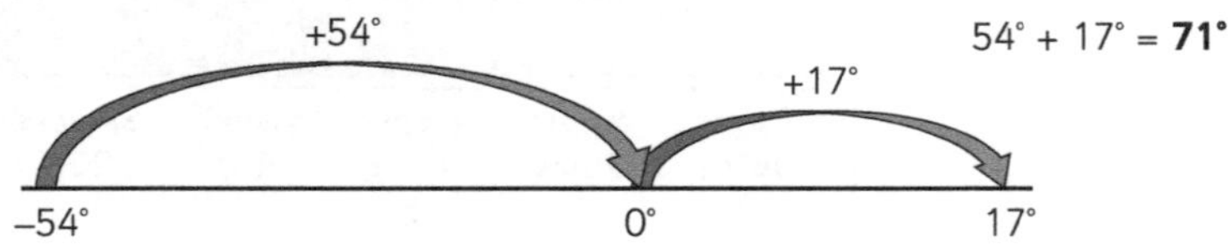 **Review** Use a numbered line (range –20 to 20) to illustrate the effect of increasing and decreasing positive and negative numbers that 'bridge' zero. If available, double-sided counters could also be used to support understanding of this concept. For example: 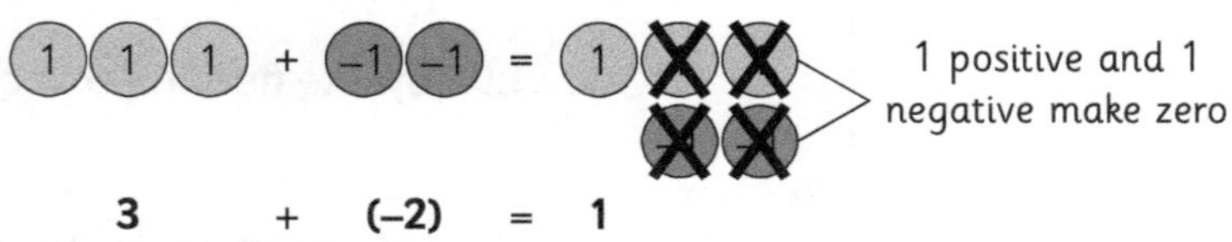
Q4b	Number & number processes MNU 2-03a	**Answer** 21°C **On track** There is evidence that pupils understand the term 'increase' and can choose an appropriate strategy to solve the problem; for example, by counting on using a number line: 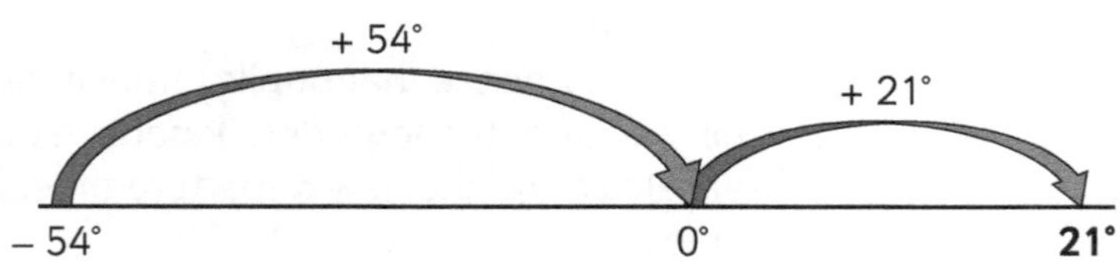 **Review** Use double sided counters and or a number line to solve contextualised addition and subtraction problems involving positive and negative numbers.

End of second level assessments: answers and marking guidance

Question	Topic	Question
Q5a	Number & number processes MNU 2-03a	**Answer** 1600 km **On track** There is evidence the pupil understands the relationship between the values in the question. i.e. 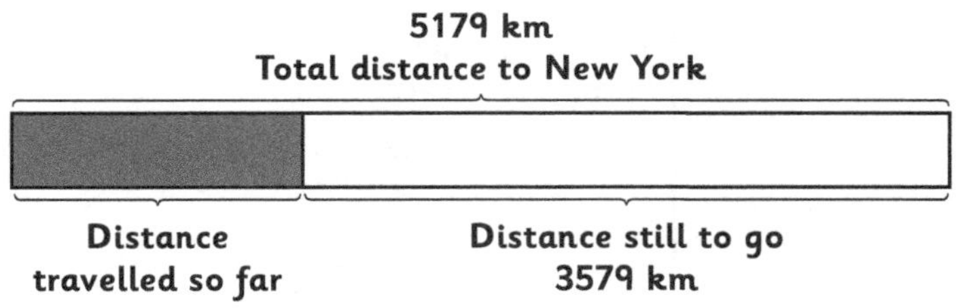 This representation allows us to see that the unknown part (the distance travelled so far) can be found by: • subtracting the known part (3579 km) from the whole (5179 km): 5179 km – 3579 km = ? • counting back from the whole (5179 km) to the known part (3579 km): 5179 km – ? = 3579 km • counting on from the known part (3579 km) to the whole (5179 km): 3579 km + ? = 5179 km 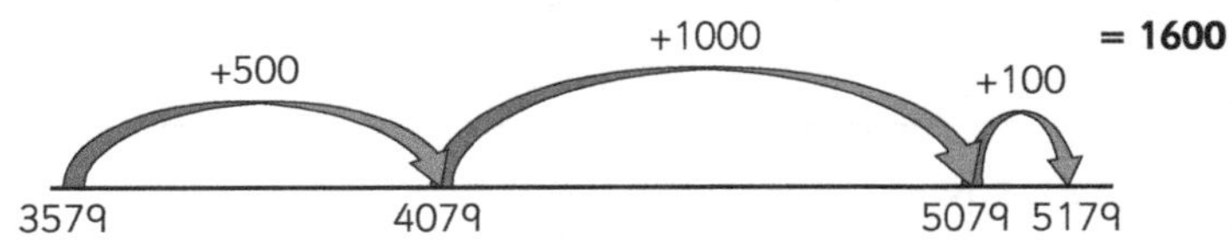 **Review** Provide opportunities to analyse and solve part-part-whole problems, within a range of contexts, where the unknown value is either the 'whole' or one of the 'parts'.
Q5b	Time MNU 2-10c	**Answer** 4 hours **On track** There is evidence the pupil understands the relationship between speed, distance and time and can apply this to analyse and solve the problem. They may represent this relationship using a bar where the whole represents both the distance travelled and time taken and each part represents the distance travelled in 1 hour (speed). Time is the unknown quantity in this problem (i.e. the number of parts is unknown) so this would require pupils to solve this problem by grouping:

End of second level assessments: answers and marking guidance

<table>
<tr><th>Question</th><th>Topic</th><th>Question</th></tr>
<tr><td></td><td></td><td>
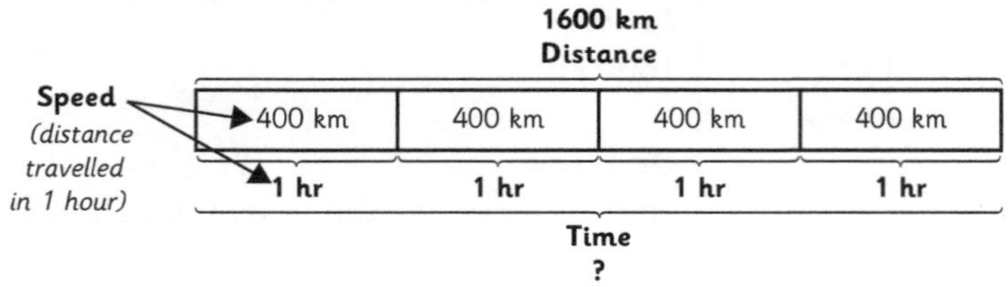

They may interpret the problem as repeated addition and choose to count on in 'jumps' of 400 km to represent the average distance travelled each hour:

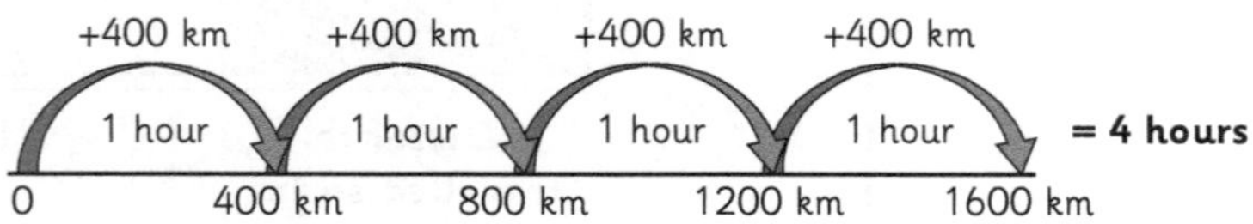

Or they may use a formula:

Distance = Speed × Time 1600 km = 400 km/h × ? '400 multiplied by what equals 1600?' OR

Time = Distance ÷ Speed ? = 1600 km ÷ 400 km/h. 'What is 1600 divided by 400?'

Review

Provide pupils with a range of simple speed, distance and time problems to analyse, and explore alternative strategies for representing and solving these. Ensure that pupils have a sound understanding of multiplication / division by multiples of 100 and 1000 and are able to use estimation to judge the reasonableness of their answer.
</td></tr>
<tr><td>Q5c</td><td>Estimation & rounding

MNU 2-01a</td><td>
Answer

Approximately 13 hours (or a reasonable estimate close to 13 hours)

On track – There is evidence the pupil understands the relationship between speed, distance and time and can apply this to analyse and solve the problem. The pupil may choose to use the answer to question 5b as a starting point and continue to count on in jumps of 400 km for the remaining 3579 km. Alternatively, they may use the total distance from question 5a and count in jumps of 400 from 0 to 5179 km. They may realise that it is easier to work out the average distance travelled in 10 hours (4000 km) and count on in 400s from there:

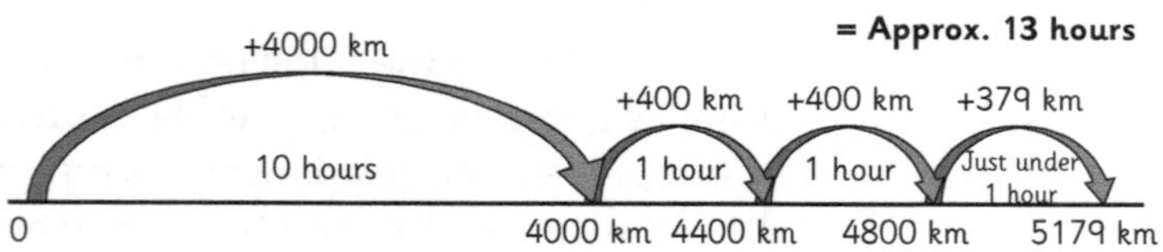

Review

Provide pupils with a range of simple speed, distance and time problems to analyse and explore alternative strategies for representing and solving these; for example, by drawing a bar model. Ensure that pupils have a sound understanding of multiplication / division by multiples of 100 and 1000 and are able to use estimation to judge the reasonableness of their answer.
</td></tr>
</table>

Question	Topic	Question
Q5d	Time MNU 2-10c	**Answer** 5:40 a.m. **On track** There is evidence that the pupil understands they need to count on 5 hours from 12:40 a.m. to account for the time difference, for example: 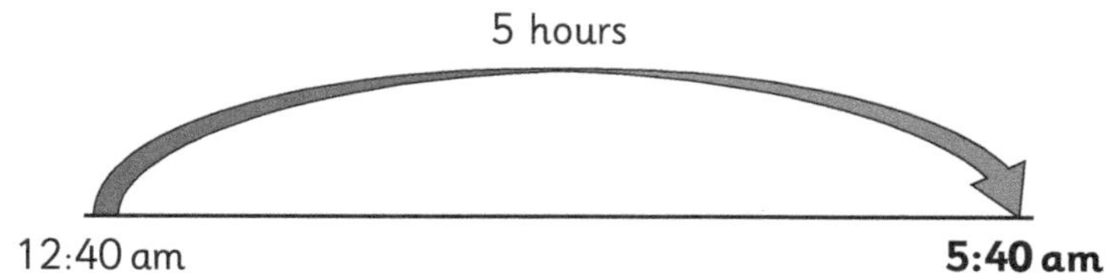 **Review** Provide opportunities to solve duration problems using both 12 and 24 hour times. Ensure that pupils can convert between 12 hr and 24 hr and have experience of working with time differences as well as duration.
Q5e	Time MNU 2-10c	**Answer** 12 hours 30 minutes **On track** There is evidence the pupil understands to count on from 17:10 to 05:40. There are different approaches that may be taken, for example: 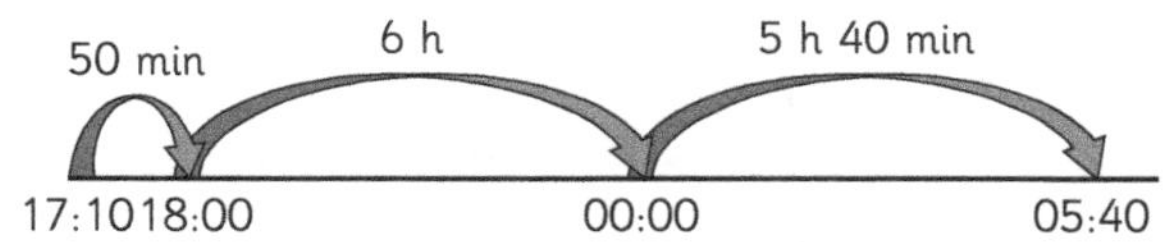 50 min + 6 h + 5 h 40 min = **12 h 30 min** **Review** Provide opportunities to calculate duration of journeys that 'bridge' a day. Use both 12 and 24 hour times. Using an empty number line can provide a support for the pupil to record and track their thinking.

Assessment 3: The escape room

Question	Topic	Question
Q1	Multiples, factors and primes MTH 2-05a	**Answer** 389614 **On track** There is evidence that pupils understand the language necessary to solve this problem (e.g. 6-digit number, factor, product) and that the position of each digit tells us its value: • 3 = 300 000 (3 hundred thousands) • 8 = 80 000 (80 thousands) • 9 = 9000 (9 thousands) • 6 = 600 (6 hundreds) • 1 = 10 (1 ten) • 4 = 4 (4 ones) **Review** Check that pupils understand the mathematical language used in the problem. Provide opportunities for them to read, write, represent and describe 6-digit numbers; for example, on place value houses.
Q2	Number & number processes MNU 2-03a	**Answer** 44 and 70 **On track** There is evidence that pupils understand the relationship between the values in the problem, i.e. 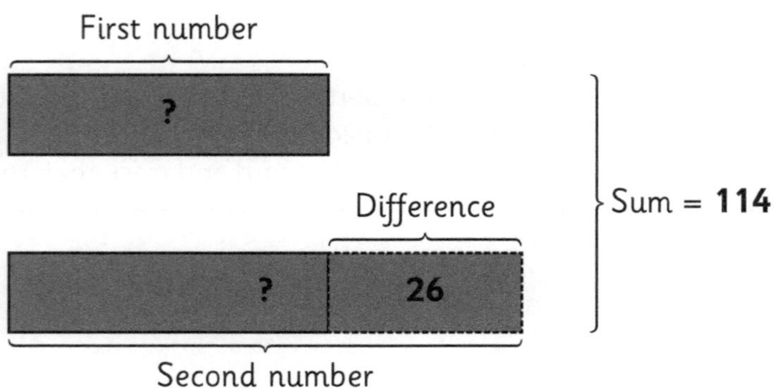 They can choose an appropriate strategy for solving the problem and justify their solution. For example, some may choose to use trial and error/guess, check and improve. Others may draw a diagram similar to the above and reason as follows: The first number can be found by: 1. subtracting 26 from 114 to get 88 2. halving 88 to get 44

<table>
<tr><th>Question</th><th>Topic</th><th>Question</th></tr>
<tr><td></td><td></td><td>The second number can be found by:
<ul><li>adding 26 to 44 to get 70; OR</li><li>finding the difference between 44 and 114</li></ul>
Review
Check the pupil understands the relevant mathematical language, e.g. sum and difference. Provide opportunities for them to solve comparison problems using visual representations; for example, bar models.</td></tr>
<tr><td>Q3</td><td>Number & number processes

MNU 2-03a</td><td>Answer
Grey (8) and white (56)

On track
There is evidence that pupils understand the relationship between the values in the problem, i.e. the white value is 7 times larger than the grey value.
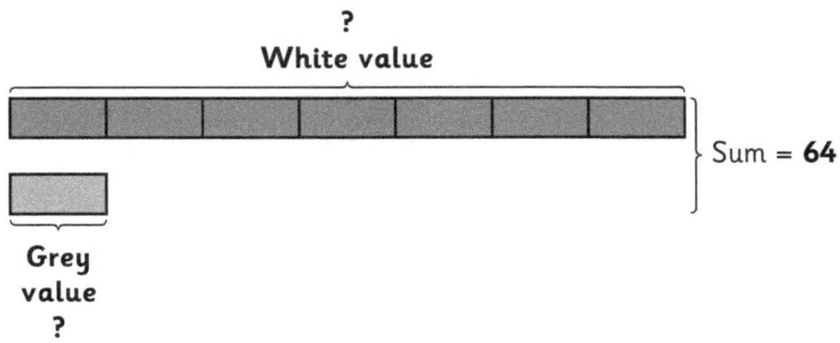

They can choose an appropriate strategy for solving the problem and justify their solution. For example, some may reason that the white value must be a multiple of 7 smaller than 64, list pairs of factors and products and add them (e.g. 6 + 42 = 48; 7 + 49 = 56; 8 + 56 = 64). Others may draw a diagram similar to the above and reason as follows:
In order to find the grey value, the sum (64) can be divided by 8. Each 'part' is worth 8, so the grey number must be 8 and the white number can be calculated by:
<ul><li>finding 7 lots of 8 (7 × 8 = 56); OR</li><li>finding the difference between 8 and 64.</li></ul>
Review
Provide opportunities for pupils to solve ratio problems using visual representations, (e.g. bar models), ensuring that they understand the relevant mathematical language.</td></tr>
</table>

End of second level assessments: answers and marking guidance

Question	Topic	Question
Q4	Expressions and equations MTH 2-15a	**Answer** Rory (18), Eva (9) and Finn (27) **On track** There is evidence that pupils understand the relationship between the values in the problem, i.e. Finn is three times older than Eva. If Eva is half Rory's age, Rory must be twice as old as Eva. 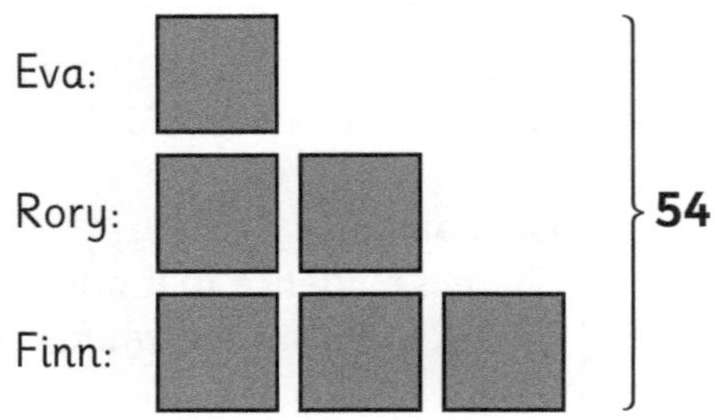 This analysis demonstrates that 6 lots of candles are needed to decorate the birthday cakes; therefore, Eva's age can be found by dividing 54 by 6 to get an **answer** of 9 candles. Rory needs 2 lots of 9 candles (18 candles) and Finn needs 3 lots of 9 candles (27 candles). The answer can be checked by adding up their ages to ensure the criteria described in the question is applicable. **Review** Provide opportunities for pupils to solve ratio problems using visual representations, e.g. bar models.
Q5a	Patterns and relationships MTH 2-13a	**Answer**  25 circles **On track** There is evidence that pupils can identify and continue the pattern as shown. **Review** Talk to the pupil about what they notice. Provide opportunities for them to create and continue similar patterns using coloured counters or drawings.

End of second level assessments: answers and marking guidance

Question	Topic	Question
Q5b	Patterns and relationships MTH 2-13a	**Answer** 36 dots **On track** There is evidence that pupils recognise the effect of adding an extra 'layer' to the pattern; for example, by sketching. Alternatively, the pupil may notice that the difference between each pair of numbers in the sequence increases by 2 each time (i.e. 5, 7, 9) therefore 11 must be added to the previous pattern of 25 dots to get an answer of 36 dots. **Review** Talk to the pupil about what they notice. Provide opportunities for them to create, continue and explain similar patterns and link these to number sequences.
Q5c	Patterns and relationships MTH 2-13a	**Answer** Add one to the pattern number and square it, so, the 5th term is $(5 + 1)^2 = 36$. The pupil may also explain the pattern as the square numbers or that the difference in the totals increases by 2 each time. **On track** There is evidence that pupils can identify: • that the difference between each term in the sequence increases by 2; OR • a link to square numbers as demonstrated: 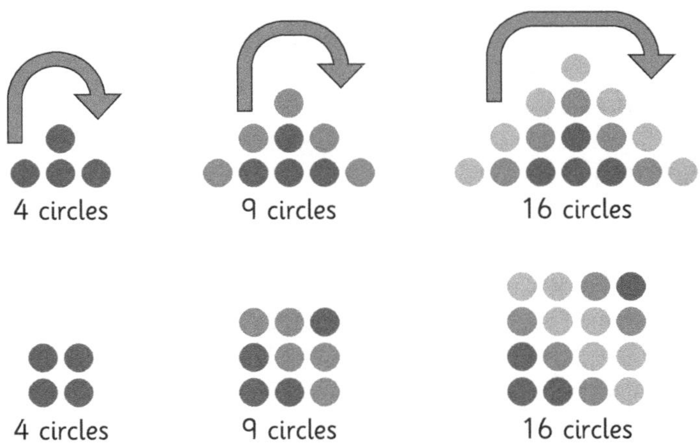 **Review** Provide opportunities for pupils to create and extend similar sequences and explain the rule.

End of second level assessments: answers and marking guidance

<table>
<tr><th>Question</th><th>Topic</th><th>Question</th></tr>
<tr><td>Q6i</td><td>Properties of 2D shapes and 3D objects

MTH 2-16b</td><td>Answer
Face B

On track
There is evidence that pupils are able to reason appropriately from the image alone; OR are able to draw and cut out the net in order to find the solution.

Review
Provide opportunities for pupils to identify and create nets for regular 3D shapes, view them from different perspectives and describe what they see.</td></tr>
<tr><td>Q6ii</td><td>Properties of 2D shapes and 3D objects

MTH 2-16b</td><td>Answer
Ability to identify any of the following 11 possibilities:
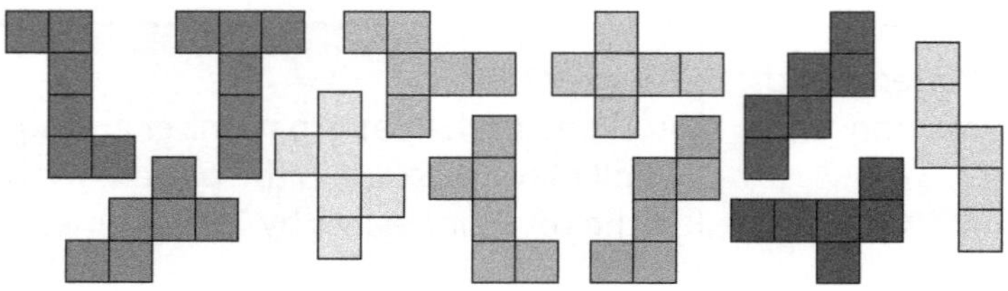
On track
There is evidence that pupils can justify why a particular configuration of six squares forms the net of a cube (whereas some configurations do not).

Review
Provide opportunities for pupils to investigate the nets of regular 3D shapes; for example, food packaging.</td></tr>
<tr><td>Q7</td><td>Measurement

MNU 2-11c</td><td>Answer
The cuboid labelled 'B' and the cube labelled 'D'

On track
There is evidence that pupils can:<ul><li>Calculate how much more water is necessary to fill the tank (15 000 ml). This can be achieved by finding the volume of the tank (100 cm × 40 cm × 30 cm = 120 000 cm³) and the volume of water (100 cm × 30 cm × 35 cm = 105 000 cm³) and calculating the difference:
OR by using the dimensions of the empty space (5 cm × 30 cm × 100 cm = 15 000 cm³ (ml).</li><li>Calculate the volume of each object using a suitable strategy. For example, the volume of the object labelled 'A' can be found by applying the following formula:<ul><li>length × height × depth</li><li>= 25 cm × 25 cm × 25 cm</li><li>= 15 625 cm³</li></ul></li></ul></td></tr>
</table>

Question	Topic	Question
		• Calculate the difference in volume between each object and the volume of empty space in the container. • Reason that the cuboid labelled 'B' and the cube labelled 'D' are the only objects with a volume less than 15 000 cm^3. **Review** Provide opportunities for pupils to calculate the volume of liquids, solids and space and make links between units, i.e. 1 ml = 1 cm^3
Q8a	Measurement MNU 2-11b	**Answer** 2.5 cm **On track** There is evidence that pupils can adopt a suitable strategy for determining the length of a small square, e.g. a small edge and a medium edge total 7.5 cm. A medium square is double the length of a small square. 7.5 cm ÷ 3 = 2.5 cm therefore the length of small square is 2.5 cm and the length of a medium square is 5 cm (2.5 cm × 2) **Review** Provide opportunities for pupils to solve, create and explain similar problems that require them to reason algebraically (using visual representations similar to those in the problem).
Q8b	Measurement MNU 2-11c	**Answer** 193.75 cm^2 **On track** There is evidence that pupils can determine the lengths of all 3 sizes of square and know how to find the area of each, using a calculator if necessary. **Review** Provide opportunities for pupils to calculate area, ensuring that they understand the link to multiplication arrays and grids.
Q8c	Measurement MNU 2-11c	**Answer** 80 cm **On track** There is evidence that pupils can determine the length of a medium side (5 cm) and a large side (10 cm). They know that perimeter means 'all the way round the edge' and can find the perimeter by adding the 'exposed' edges. **Review** Check that pupils understand the meaning of the term 'perimeter' and can distinguish between perimeter and area. Provide opportunities for pupils to calculate the perimeter of regular and irregular shapes.

Question	Topic	Question
Q8d	Measurement MNU 2-11c	**Answer** A medium square with only one side adjacent to the original shape, for example: 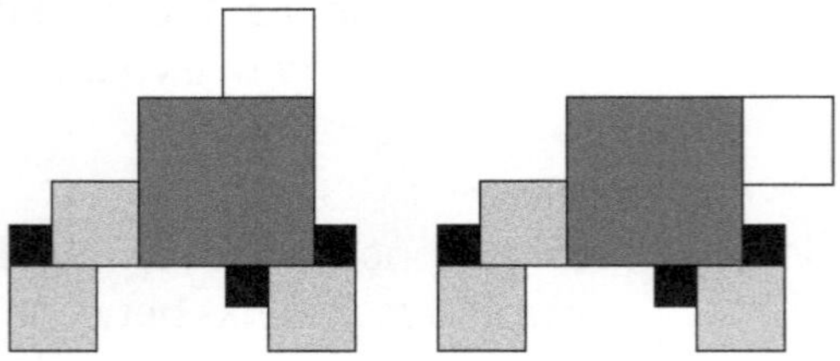 **On track** There is evidence that pupils are able to reason that the only square that can be used to do this is a medium one in a position where only one of its sides is adjacent to the sides of the existing shape. Alternatively, accept a small square being added where only a corner meets another corner, i.e. no side of the added square is adjacent to any side of the existing shape. **Review** Provide opportunities for pupils to analyse, solve and create similar problems.

End of second level assessments: answers and marking guidance

Assessment 4: The carnival

Question	Topic	Question
Q1a	Time MNU 2-10a	**Answer** 14:04 **On track** There is evidence that pupils can convert between analogue and digital time and can read and interpret the timetable to determine the correct train to reach Queen Street Station by 3 p.m. (15:00). **Review** Check that the pupil can convert between digital and analogue time. Provide opportunities for pupils to read, interpret and create timetables of varying layouts for a range of contexts.
Q1b	Time MNU 2-10a	**Answer** 13:37 **On track** There is evidence that pupils can reason about the problem and calculate Amman's departure time using a suitable strategy, for example: 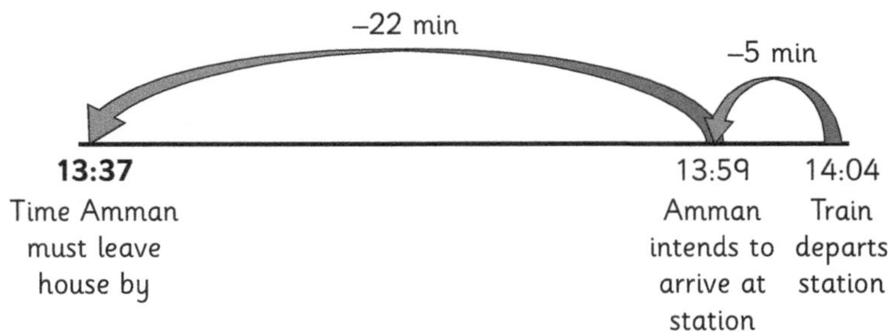 **Review** Check that the pupil can identify, and correctly use, an appropriate strategy for calculating duration of time, with an emphasis on situations that require them to bridge an hour, or hours. Provide opportunities for pupils to analyse and solve time duration problems.
Q2a	Angle, symmetry and transformation MTH 2-17a	**Answer** 30° **On track** There is evidence that pupils are able to choose and apply a strategy to work out the size of the angle; for example, they may divide 360° by 12 or 90° by 3. **Review** Ensure pupils know that complementary angles add to 90°, supplementary angles add to 180° and that a 'full turn' measures 360°. If necessary, provide practice in dividing multiples of 10 using concrete materials.

End of second level assessments: answers and marking guidance

Question	Topic	Question
Q2b	Angle, symmetry and transformation MTH 2-17a	**Answer** acute angle **On track** There is evidence that pupils understand that the hour hand will have moved to halfway between the 3 and the 4, thus making the resultant angle acute. **Review** Provide opportunities for pupils to identify, name and create different types of angle.
Q2c	Angle, symmetry and transformation MTH 2-17a	**Answer** 75° **On track** There is evidence that pupils: i. understand that the hands of the clock will be positioned as follows:  ii. know that a right angle is 90 degrees iii. can identify and use an appropriate strategy to solve the problem; for example, by calculating the angle between each numeral on the clock face (perhaps by dividing 90° by 3) and then halving this in order to calculate the angle between the hour hand and the numeral 4. The size of the angle is therefore 30° + 30° + 15° **Review** Provide opportunities for pupils to identify, measure and calculate different angles.

End of second level assessments: answers and marking guidance

Question	Topic	Question
Q3	Money MNU 2-09a	**Answer** 2 family tickets and 1 child ticket: £86.33 **On track** There is evidence that pupils can identify and compare the various options they have for buying tickets i.e. • 3 adult tickets (3 × £12.49) and 6 child tickets (£8.95) • 1 group ticket (9 × £9.85) • 2 family tickets (2 × £38.69) and 1 child ticket (£8.95) **On track** Provide opportunities for pupils to analyse and solve multi-step, comparison problems involving money.
Q4a	Angle, symmetry and transformation MTH 2-17a	**Answer** Answer depends on route taken between the ticket office and waltzers (and the scale at which the resource is printed), but in the End of Level assessment, it is approximately 6 cm, so 30 m; in the downloadable resource, it is approximately 9 cm, so 45 m. **On track** There is evidence that pupils can measure the distance from the ticket office to the Waltzer on the floorplan and calculate the actual distance using the scale 1 cm = 5 m **Review** Provide opportunities for pupils to measure and calculate distances using simple scales on maps and plans.
Q4b	Angle, symmetry and transformation MTH 2-17a	**Answer** Walk through the entrance to the left of the ticket office and head north-west (or on a bearing of 315°). Accept alternative directions that describe a suitable route around the various attractions along the way. **On track** There is evidence that pupils know the 8-point compass, can use and apply appropriate directional language (e.g. degrees, bearing, etc.). **Review** Check that the pupil can make the link between compass points and angles. Provide opportunities for them to follow and give directions using the 8-point compass and its equivalent angles.

End of second level assessments: answers and marking guidance

Question	Topic	Question
Q5a	Fractions, decimal fractions and percentages MNU 2-07a	**Answer** 6 rides **On track** There is evidence that pupils can analyse and solve the problem using an appropriate strategy, for example: 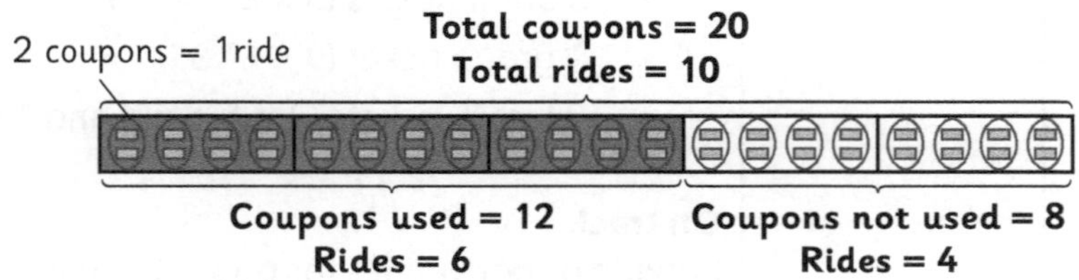 **Review** Provide opportunities for pupils to analyse and solve problems that involve finding the fraction of a value using visual representations, e.g. bar models.
Q5b	Fractions, decimal fractions and percentages MNU 2-07a	**Answer** 24 coupons each **On track** There is evidence that pupils can use the information from the previous question to deduce that Amman did not use 8 of his 20 coupons. The remaining 8 were shared equally between Isla and Nuria, giving the girls an extra 4 coupons each. **Review** Provide opportunities for pupils to analyse and solve problems that involve finding the fraction of a value using visual representations, e.g. bar models.
Q5c	Fractions, decimal fractions and percentages MNU 2-07a	**Answer** one third of her coupons **On track** There is evidence that pupils can analyse and solve the problem using an appropriate strategy, for example: Isla uses 16 out of 24 coupons, which can be represented as follows: 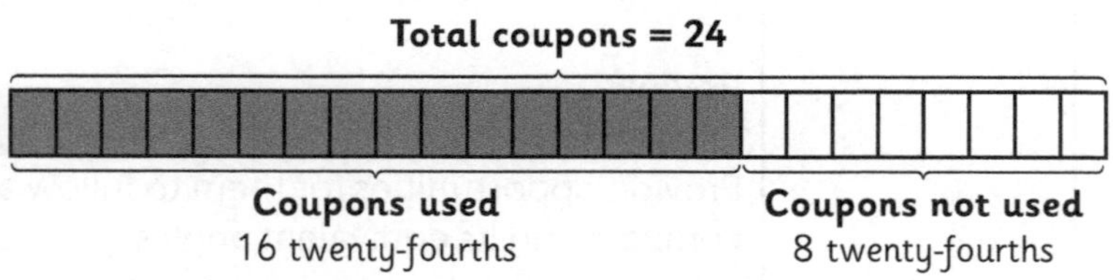

Question	Topic	Question
		Which can be simplified as: 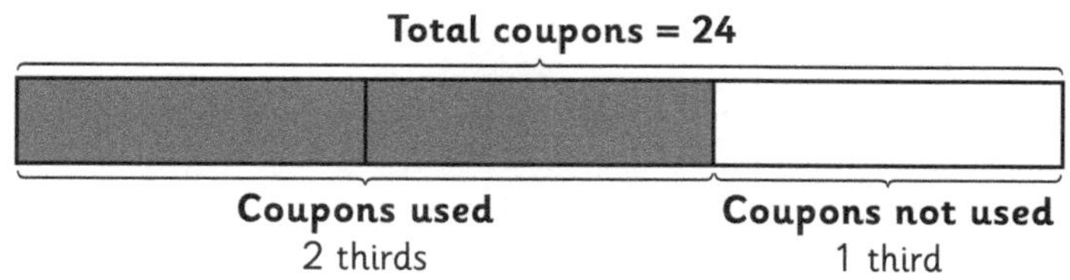 **Review** Check that the pupil understands the term 'simplest form'. Provide opportunities for them to analyse and simplify fractions using visual representations, e.g. bar models.
Q5d	Fractions, decimal fractions and percentages MNU 2-07a	**Answer** 9 rides (18 coupons) **On track** There is evidence that pupils can analyse and solve the problem using an appropriate strategy, for example: 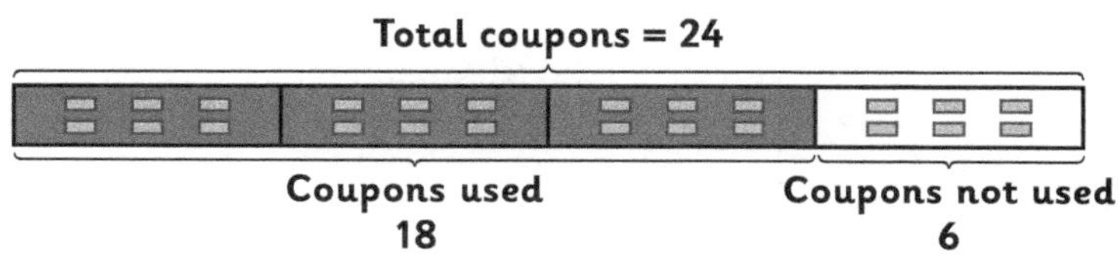 The total number of coupons can be shared equally between each quarter 'part'. The coupons Nuria used can be calculated by adding up all of the coupons relating to 3 of these four equal parts. **Review** Provide opportunities for pupils to analyse and solve problems that involve finding the fraction of a value using visual representations, e.g. bar models.
Q6a	Money MNU 2-09a	**Answer** A burger and cola **On track** There is evidence that pupils can identify and apply a suitable strategy; for example, they may use 'trial and improvement' to randomly select and total two items, using the **answer** to refine their choices. They may use estimation and rounding to find approximate totals and immediately rule out certain items. Alternatively, they may reason that the 'pennies' must add together to 'give 1 in the hundredths place' (in this case 6p + 5p). **Review** Provide opportunities for pupils to analyse and solve part-part-whole problems in a range of contexts.

End of second level assessments: answers and marking guidance

Question	Topic	Question
Q6b	Money MNU 2-09a	**Answer** Pizza **On track** There is evidence that the child can analyse the problem and apply a suitable strategy for finding the amount that Amman spent, for example: 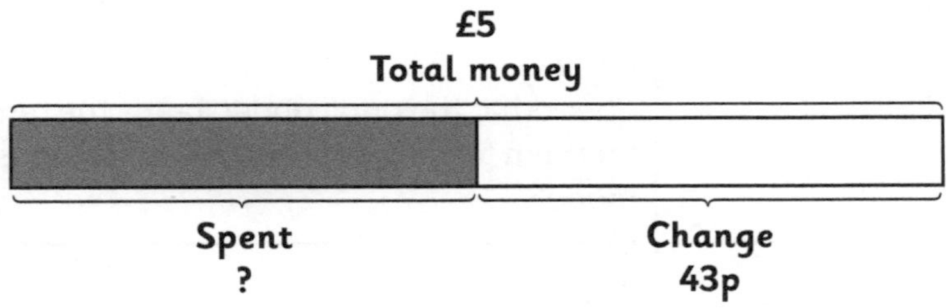 This representation allows us to see that the unknown part can be found by, for example, subtracting 43p from £5: 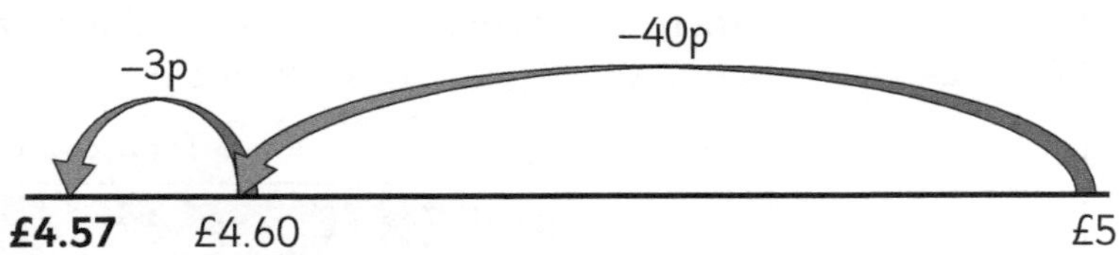 **Review** Provide opportunities for pupils to analyse and solve part-part-whole problems in a range of contexts.
Q6c	Money MNU 2-09a	**Answer** Nuggets and a coffee **On track** There is evidence that the child can analyse the problem and apply a suitable strategy for finding the amount that Nuria spent, for example: 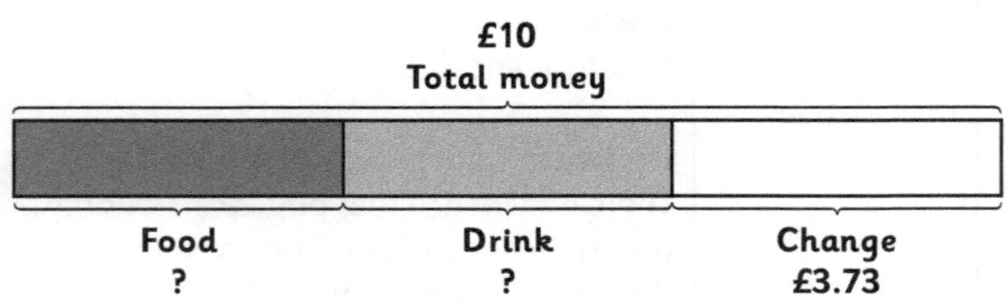 This representation allows us to see that the unknown part can be found by, for example, subtracting £3.73 from £10: 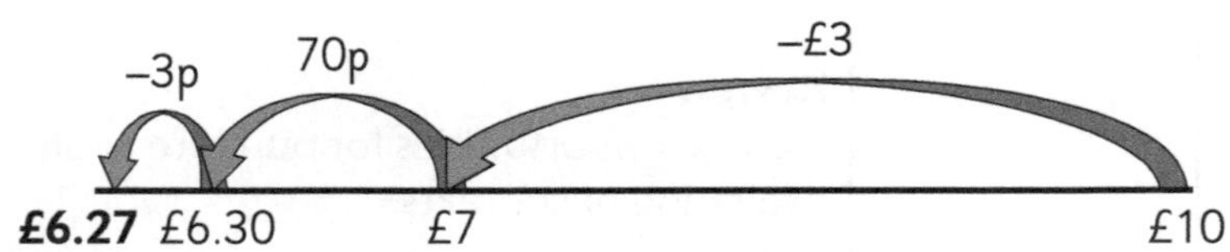

End of second level assessments: answers and marking guidance

Question	Topic	Question
		From here, pupils may use trial and improvement to find how the £6.27 was split between two items. Alternatively, they may reason that the 'pennies' must add together to 'give 7 in the hundredths place' (8p + 9p) **Review** Provide opportunities for pupils to analyse and solve multi-step problems in a range of contexts.
Q6d	Money MNU 2-09a	**Answer** A hot dog and water **On track** There is evidence that the child can analyse the problem and apply a suitable strategy for finding the items that Isla bought; for example, they may decide to represent the 'three times as much' ratio between the cost for food and drink using a bar model to compare them: 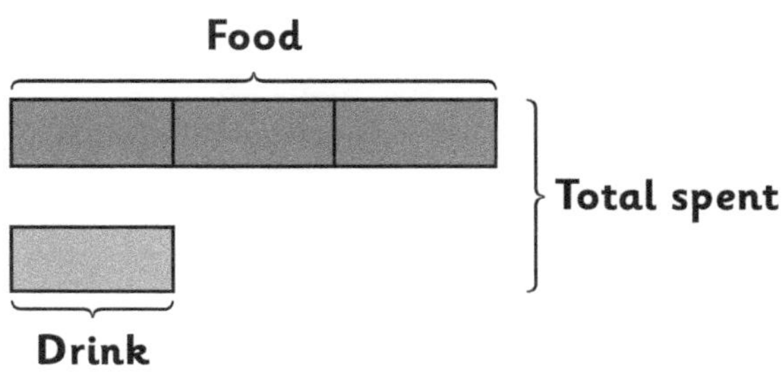 From here, pupils may reason that the price of the food must be divisible by 3, eliminate all but the hot dog and fries, then use trial and improvement to find a drink that is one third of the price of these items. Alternatively, they may identify items where there is a multiplicative relationship between the number of pennies that will result in 'the hundredths place' for the cost of the food being three times more than the cost of the drink, i.e. 7p and 21p. **Review** Provide opportunities for pupils to analyse and solve ratio problems in a range of contexts.
Q7a	Ideas of chance and uncertainty MNU 2-22a	**Answer** Finlay should call 'grey' because grey covers half (50%) of the wheel's surface. **On track** There is evidence that pupils can reason that Finlay should call grey because there is a 50% chance of being correct as opposed to black (1 in 6 chance) or white (1 in 3 chance). **Review** Check that pupils can interpret a pie chart and apportion a fraction and/or percentage to each sector. Provide opportunities for pupils to analyse and solve problems involving chance and uncertainty.

End of second level assessments: answers and marking guidance

Question	Topic	Question
Q7b	Ideas of chance and uncertainty MNU 2-22a	**Answer** No impact **On track** There is evidence that pupils can reason that the previous 10 results have no bearing on Finlay's result. He still has a 50% chance if he sticks with a call of grey so he should stick with his original decision. **Review** Provide opportunities for pupils to analyse and solve problems involving chance and uncertainty.

Assessment 5: The festival

Question	Topic	Question
Q1a	Angle, symmetry and transformation MTH 2-17a	**Answer** 45° **On track** There is evidence that pupils understand the link between compass points and bearings, i.e. north-east = 45° **Review** Check that pupils are familiar with the 8-point compass, understand that a full turn equals 360 degrees and that bearings are always measured clockwise from north. Provide opportunities for them to convert compass points into bearings and vice versa.
Q1b	Angle, symmetry and transformation MTH 2-17a	**Answer** (5,4) **On track** There is evidence that pupils can plot and record the location of a point using coordinate notation. **Review** Provide opportunities for pupils to describe, plot and record coordinate points in the form (x,y).
Q1c	Angle, symmetry and transformation MTH 2-17a	**Answer** cycle hire **On track** There is evidence that pupils can interpret directional information (bearing of 000° / west) and apply their understanding of scale (i.e. each square represents 1 km^2 so each side represents 1 km). **Review** Provide opportunities for pupils to use and apply directional language to interpret plans and maps with simple scales.
Q1d	Angle, symmetry and transformation MTH 2-17a	**Answer** Multiple answers are possible; for example, 'travel south-east (on a bearing of 135°) for approximately 5 km.' **On track** There is evidence that pupils can locate the information centre and castle on the map and provide relevant directions from point (9,16) to point (14,11).

End of second level assessments: answers and marking guidance

Question	Topic	Question
		Review Check that pupils are familiar with the 8-point compass, understand that a full turn equals 360 degrees and that bearings are always measured clockwise from north. Provide opportunities for pupils to apply their understanding of scale and directional language (compass points and bearings).
Q1e	Angle, symmetry and transformation MTH 2-17a	**Answer** (14,5) **On track** There is evidence that pupils can locate the garden and horse riding on the map and use their understanding of symmetry to locate and record the coordinates of the picnic site. **Review** Provide opportunities for pupils to use coordinate notation to describe, plot and record points. Ensure that pupils are able to identify symmetrical shapes and complete symmetrical pictures with two or more lines of symmetry.
Q2a	Measurement MNU 2-11c	**Answer** 50 m (1 m by 24 m) **On track** There is evidence that pupils understand the terms area and perimeter, can distinguish between them and can apply their understanding to determine possible dimensions of the rectangular pitch. **Review** Check the pupil's understanding of the terms area and perimeter, and the relationship between them. Provide opportunities for pupils to calculate the perimeter of different sized rectangles. Determine whether they are relying on a strategy like 'guess, check and improve' or can reason that making the longest rectangle possible (i.e. 1 m by 24 m) will result in the greatest possible perimeter.
Q2b	Measurement MNU 2-11c	**Answer** 20 m (4 m by 6 m) **On track** As in Q2a. Is the child able to reason that a rectangle (or array) produced using dimensions closest together in length (i.e. 4 m by 6 m) will result in the minimum perimeter? **Review** Check the pupil's understanding of the terms area and perimeter, and the relationship between them. Provide opportunities for pupils to calculate the perimeter of different sized rectangles and solve factorisation problems using arrays.

End of second level assessments: answers and marking guidance

Question	Topic	Question
Q3ai	Measurement MNU 2-11b	**Answer** 4250 g or 4.25 kg or 4 kg 250 g **On track** There is evidence that pupils can read and interpret scales accurately, appreciating that they must first determine the value of each increment (and that this can be different on different sets of scales). **Review** Check that the pupil has not just made an assumption as to the value of each increment and applied this value either correctly or incorrectly. Provide opportunities for them to read and interpret scales relating to mass and convert between the different units, i.e. grams and kilograms.
Q3aii	Measurement MNU 2-11b	**On track** 1850 g or 1.85 kg or 1 kg 850 g **On track** As in Q3a. **Review** As in Q3a.
Q3b	Measurement MNU 2-11b	**Answer** 35 litres (or 35 000 ml) **On track** There is evidence that pupils can read and interpret the scale accurately and can multiply decimal numbers by 10. **Review** Provide opportunities for pupils to read and interpret scales relating to capacity and convert between different units, e.g. 500 ml = ½ l. Check the pupil's ability to multiply whole and decimal numbers by 10 and provide practice if required.

End of second level assessments: answers and marking guidance

<table>
<tr><th>Question</th><th>Topic</th><th>Question</th></tr>
<tr><td>Q4</td><td>Money

MNU 2-09c</td><td>Answer
51 cups

On track
There is evidence that pupils can analyse the problem and apply a suitable strategy for calculating the total cost of the ingredients (£76.28). From here, it will be necessary to identify how many cups need to be sold to take them over this cost. Some children may attempt to solve this by dividing £76.28 by £1.50. Not only will this be tricky to calculate but will give an answer of 50.8533 which some children may have difficulty interpreting. A more effective strategy would be to count how many lots of £1.50 will be necessary to take them over £76.28. Rather than doing this one at a time, it would be more efficient to perform this in multiples of 10; for example, 10 cups sold returns £15; 20 cups returns £30; 40 cups returns £60; 50 cups returns £75. From here, 1 more cup needs to be sold to take the total income above £76.28 (i.e. £76.50). Therefore, the minimum number of cups needed to be sold would be 51 cups resulting in a profit of 22p.

Review
Check that the pupil can choose, and correctly use, appropriate strategies for multiplying whole numbers by multiples of ten and adding numbers with up to two decimal places. Provide opportunities for pupils to analyse and solve:
• Multi-step problems
• Division problems using grouping</td></tr>
<tr><td>Q5a</td><td>Money

MNU 2-09c</td><td>Answer
Profit

On track
There is evidence that pupils understand the terms profit and loss (i.e. Total Income > Total Cost results in a profit; Total Income < Total Cost results in a loss) Some may perform a calculation to determine whether a profit / loss is made; others may be able to reason based on their answer to Q4, i.e. 'They only need to sell 51 cups to make a profit; 70 cups is 19 more than they needed to sell; therefore, they must make a profit.'

Review
Check that the pupil understands the terms profit and loss. Provide opportunities for them to analyse and solve:
• Profit / loss problems
• Multiplication problems involving multiples of ten</td></tr>
</table>

Question	Topic	Question
Q5b	Money MNU 2-09c	**Answer** – £28.72 **On track** There is evidence that pupils understand the terms profit and loss and can choose, and correctly use, an appropriate strategy for calculating the difference between the two amounts; for example, by counting on, counting back or subtraction. **Review** Check that the pupil understands the terms profit and loss. Provide opportunities for them to calculate profit and loss.

End of second level assessments: answers and marking guidance

Assessment 6: The activity day

Question	Topic	Question
Q1	Number and number processes MNU 2-02a	**Answer** – 1st: 130 000 (Nuria); 2nd: 125 750 (Finlay); 3rd: 117 500 (Amman); 4th: 98 250 (Isla) **On track** There is evidence that pupils can read and write 5- and 6-digit numbers, in words and numerals, and can choose a strategy for ordering them, i.e. by comparing the number of digits and their values. **Review** Provide opportunities for pupils to: • Write 5- and 6-digit numbers presented in words in numerals, and vice versa • Compare and order numbers with up to 6 digits
Q2	Number and number processes MNU 2-02a	**Answer** 2a: Finlay: 104 000; Amman: 55 250 **Answer** 2b: 353 500 (Finlay + Amman + Isla + Nuria = 104 000 + 55 250 + 83 250 + 111 000) **On track** There is evidence that pupils can calculate each of the total scores appropriately by applying their understanding of multiplying, adding and subtracting multiples of 10, 100, 1000 and 10 000. **Review** Check that pupils can choose and use appropriate strategies for adding, subtracting and multiplying numbers with up to six digits, including multiples of 100 and 1000. Provide opportunities for them to: • Demonstrate their understanding of place value by partitioning numbers with up to 6 digits • Analyse and solve multi-step word problems involving multiplication, division, addition and subtraction of multiples of 10, 100, etc.
Q3a	Number and number processes MNU 2-03b	**Answer** 1st: 35.05 seconds (Ali) 2nd: 35.082 seconds (Nuria) 3rd: 35.28 seconds (Finlay) 4th: 35.8 seconds (Petra) 5th: 36 seconds (Amman) 6th: 36.021 (Eva) 7th: 36.12 seconds (Max) 8th: 36.2 seconds (Isla)

End of second level assessments: answers and marking guidance

<table>
<tr><th>Question</th><th>Topic</th><th>Question</th></tr>
<tr><td></td><td></td><td>

On track

There is evidence that pupils:

- Can read, write, compare and order decimal numbers with up to three decimal places
- Understand that the smallest number equates to the fastest time and vice-versa

Review

Check that pupils can read, write, build and describe numbers with up to three decimal places and provide opportunities for them to compare decimal numbers using concrete materials.

Some children may incorrectly assume that, for example, 36.021 seconds is greater than 36.12 seconds (because they apply whole number logic, i.e. that 21 is greater than 12). It is important that pupils appreciate the scale of each digit by using both concrete and pictorial representations. Dienes blocks can be used, or tenths and hundredths produced by the pupils themselves; for example, by cutting thick card to scale:

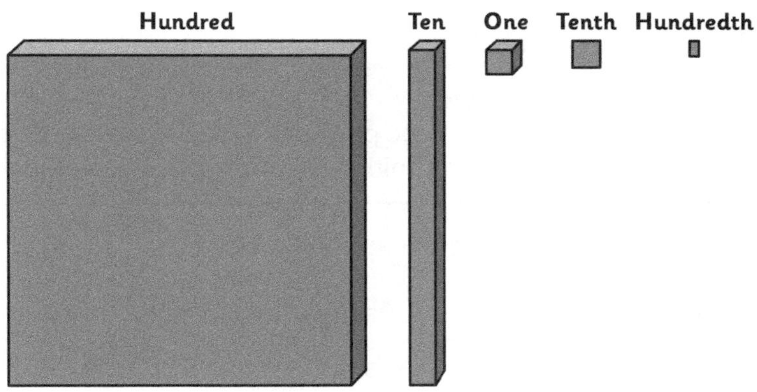

</td></tr>
<tr><td>Q3b</td><td>Number and number processes

MNU 2-03b</td><td>

Answer

0.92 seconds

On track

There is evidence that pupils understand how to find the difference between a value to 2 decimal places; and another to 1 decimal place, for example, by counting on:

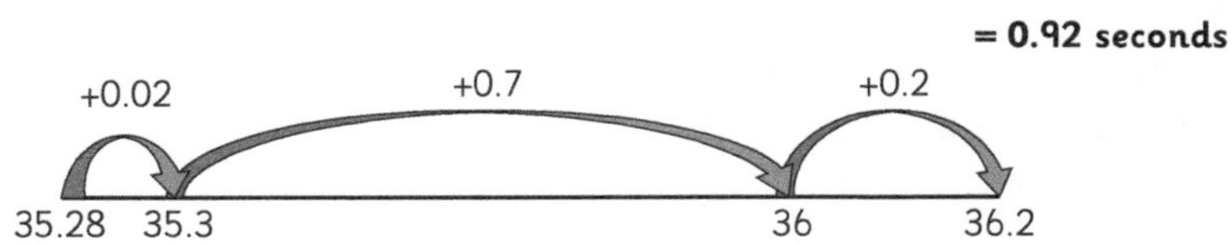

Review

Provide opportunities for pupils to compare decimal values (particularly values with a different number of decimal places) using concrete materials.

</td></tr>
</table>

End of second level assessments: answers and marking guidance

Question	Topic	Question
Q3c	Number and number processes MNU 2-03b	**Answer** i) 352.8 seconds ii) 3528 seconds iii) 35 280 seconds **On track** There is evidence that pupils understand that when, for example, multiplying by 10, each digit is getting 10 times bigger. For example, 8 hundredths is scaled up to 8 tenths. **Review** Provide opportunities for pupils to multiply decimal numbers by 10, 100 & 1000 using concrete materials and visual representations.
Q4a	Time MNU 2-10a	**Answer** 6 minutes **On track** There is evidence that pupils understand that time has 60 as its base (rather than 10) and apply an appropriate strategy for determining the correct number of minutes; for example, by counting on from 0 to 360 in 60s, or by dividing 360 by 60. **Review** Check the pupil's ability to convert between hours and minutes. Provide opportunities for pupils to solve problems involving time conversions.
Q4b	Number and number processes MNU 2-03b	**Answer** Finlay: 34.61 secs; Isla: 33.908 secs; Nuria: 36 secs; Amman: 34.325 secs; Eva: 33.77 secs; Max: 36.51 secs; Petra: 35.868 secs; Ali: 34.165 secs **On track** There is evidence that pupils understand that when dividing by 10, each digit is getting 10 times smaller so, for example, 8 tenths is scaled down to 8 hundredths. **Review** Provide opportunities for pupils to divide decimal numbers by 10, 100 and 1000 using concrete materials and visual representations.
Q4c	Number and number processes MNU 2-03b	**Answer** 380.71 seconds **On track** There is evidence that pupils can analyse and solve this problem by, for example, by representing the relationships between the values using a bar model:

Question	Topic	Question
	Fractions, decimal fractions and percentages MNU 2-07b	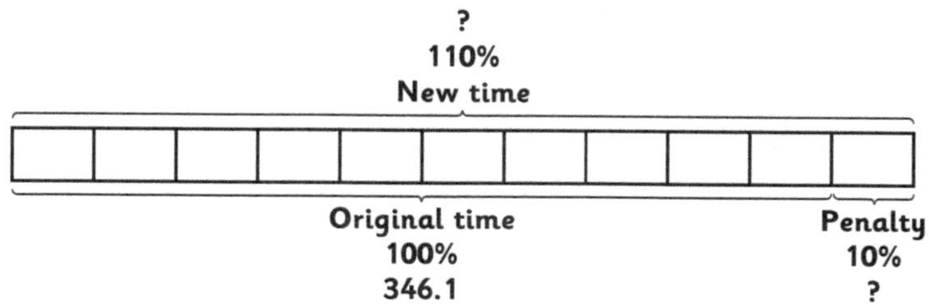 The pupil appreciates that 10% can be calculated by finding one tenth of Finlay's time (or dividing his time by 10) and that this needs to be **added** to Finlay's time. **Review** Ensure the pupil understands that a 10% penalty means that 10% will be added to his time, can equate 10% with 'one tenth' and can choose and use a suitable strategy for finding $\frac{1}{10}$ of Finlay's time. Provide them with opportunities to analyse, represent and solve percentage problems; for example, by using a bar model.
Q5a	Fractions, decimal fractions and percentages MTH 2-07c	**Answer** Finlay (3 quarters is greater than 5 eighths because 3 quarters is equal to 6 eighths) **On track** There is evidence that pupils can compare fractions with different denominators and justify their answer. For example, Finlay and Amman's distances can be compared using a bar model to demonstrate that three quarters is equivalent to six eighths (which can be compared directly with five eighths): 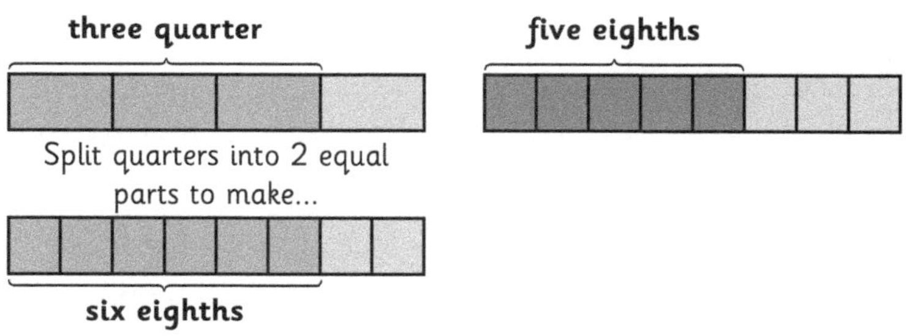 **Review** Check that pupils can compare fractions with the same denominator. Provide opportunities for pupils to compare fractions with different denominators using visual representations.

End of second level assessments: answers and marking guidance

Question	Topic	Question
Q5b	Fractions, decimal fractions and percentages MTH 2-07c	**Answer** Nuria (2 thirds is greater than one half because 4 sixths is greater than 3 sixths). **On track** There is evidence that pupils can compare fractions with different denominators and justify their answer. For example, Isla and Nuria's distances could be compared using a bar model to demonstrate that one half is equivalent to three sixths and two thirds is equivalent to four sixths. Once both fractions have been converted to sixths, a direct comparison can be made: 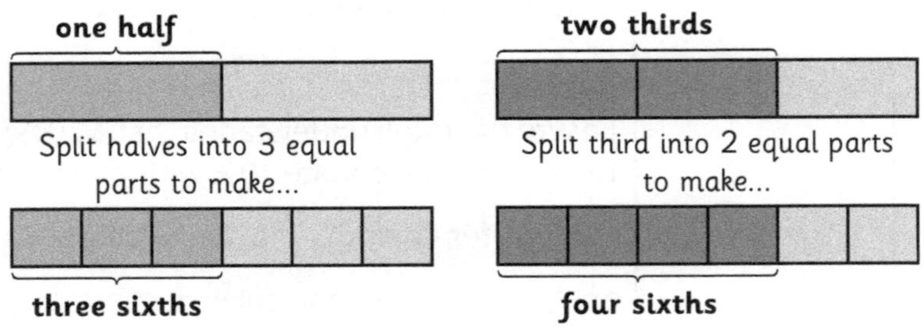 Pupils may reason that Isla failed to complete one half of the track whereas Nuria only failed to complete one third of the track, therefore Nuria travelled a greater distance. NB. The most efficient way to compare one half and two thirds using equivalence is to transform them both into sixths. However, the same fractions could also be compared directly if transformed into twelfths, eighteenths, twenty-fourths, etc. **Review** Provide opportunities for pupils to compare fractions with different denominators, using visual representations. Encourage them to look for and comment on 'common factors' and find and make equivalent fractions.
Q5c	Fractions, decimal fractions and percentages MTH 2-07c	**Answer** Finlay (3 quarters is greater than 2 thirds because 9 twelfths is greater than 8 twelfths) **On track** There is evidence that pupils appreciate that, in order to work out who was fastest **overall,** they need only compare Finlay and Nuria's times, and realise that this means comparing $\frac{3}{4}$ (fastest boy) and $\frac{2}{3}$ (fastest girl) by finding a common denominator. This could be done using a bar model to demonstrate that three quarters is equivalent to nine twelfths and two thirds is equivalent to eight twelfths. Once both fractions have been converted to twelfths, a direct comparison can be made:

<table>
<tr><th>Question</th><th>Topic</th><th>Question</th></tr>
<tr><td></td><td></td><td>
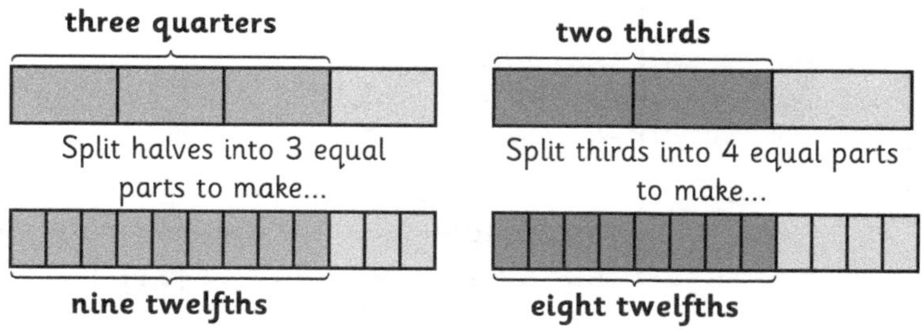

Some pupils may reason that Nuria failed to complete one third of the track whereas Finlay only failed to complete one quarter of the track; therefore, Finlay travelled a greater distance.

Review
Check that pupils can represent fractions using visual representations and compare them by looking for common factors.
</td></tr>
<tr><td>Q6a</td><td>Number and Number Processes

MNU 2-03a</td><td>
Answer
146 yards

On track
There is evidence that the child can analyse the problem and apply a suitable strategy for finding the distance to the flag. For example:
The problem can be analysed as follows:

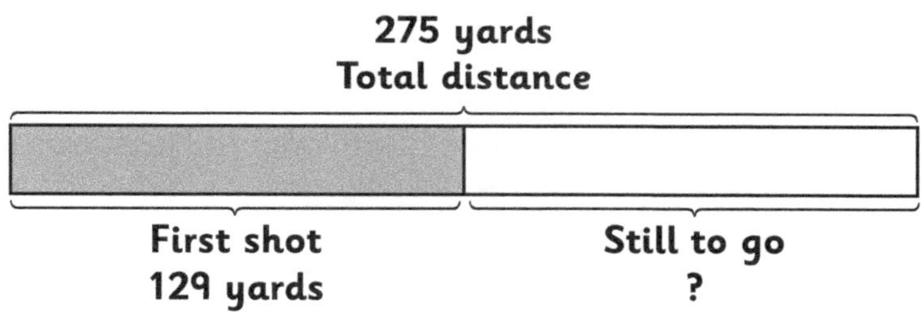

This representation allows us to see that the unknown part can be found by, for example, counting on from 129 up to 275:

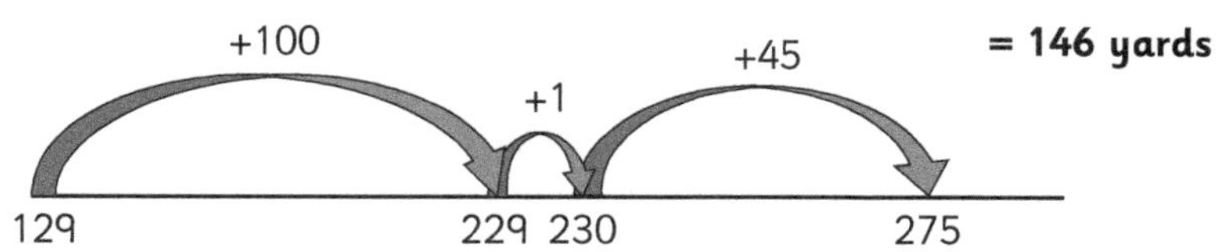

Some pupils may choose to count back from 275, down to 129; others may choose to subtract 129 from 275.

Review
Provide pupils with opportunities to analyse and solve part-part-whole problems where the unknown value is either the whole or one of the parts, using a range of contexts.
</td></tr>
</table>

End of second level assessments: answers and marking guidance

<table>
<tr><th>Question</th><th>Topic</th><th>Question</th></tr>
<tr><td>Q6b</td><td>Number and Number Processes

MNU 2-03a</td><td>

Answer
188 yards

On track
There is evidence that the child can analyse the problem and apply a suitable strategy for finding the distance to the flag. For example:
The problem can be analysed as follows:
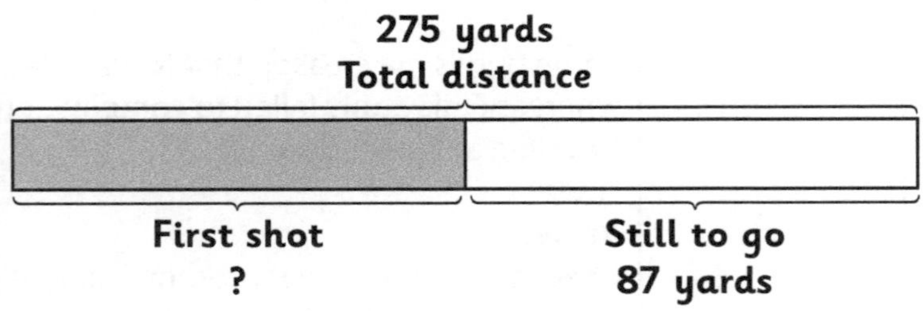

This representation allows us to see that the unknown part can be found by, for example, subtracting 87 from 275:
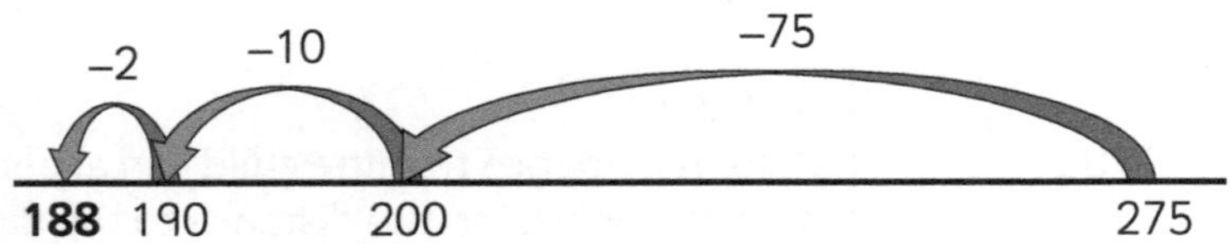

Some pupils may choose to count back from 275, down to 87; others may choose to count on from 87 up to 275.

Review
Provide pupils with opportunities to analyse and solve part-part-whole problems where the unknown value is either the whole or one of the parts, using a range of contexts.
</td></tr>
<tr><td>Q6c</td><td>Angle, symmetry and transformation

MTH 2-17a</td><td>

Answer
38°

On track
There is evidence that pupils understand that the vertical line makes a 90° angle with the ground and that the missing angle can be found by finding the difference between 52° and 90°.

Review
Provide pupils with opportunities to analyse and solve problems involving complementary and supplementary angles.
</td></tr>
</table>